Inspection and Measurement in Manufacturing

Other SME Books Authored or Edited by
William Winchell

*Continuous Quality Improvement: A
Manufacturing Professional's Guide*

*Continuous Improvement in Action: Eight
Original In-depth Case Studies*

*Realistic Cost Estimating for Manufacturing,
2nd Edition*

Inspection and Measurement in Manufacturing

keys to process planning and improvement

William Winchell

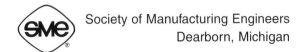

Society of Manufacturing Engineers
Dearborn, Michigan

Library of Congress Catalog Card Number: 96-067890
International Standard Book Number: 0-87263-474-4

Additional copies may be obtained by contacting:

Society of Manufacturing Engineers
Customer Service
One SME Drive
Dearborn, MI 48121
1-800-733-4763

SME staff who participated in producing this book:

Larry Binstock, Senior Editor
Rosemary Csizmadia, Operations Administrator
Dorothy Wylo, Production Assistant
Frances Kania, Production Assistant

Cover design by: Margaret Novak

Printed in the United States of America

"When you can measure what you are speaking about and express it in numbers, you know something about it; and when you cannot measure it, when you cannot express it in numbers, your knowledge is of a meager and unsatisfactory kind."

— Lord William Thomson Kelvin

Contents

Part 2—MEASUREMENT

Preface

Inspection and measurement have undergone much change. Now they are handled in the manufacturing area instead of the quality department in many companies. No longer "standalone" in nature, inspection and measurement are being addressed and planned up-front by product development teams. They are also now orchestrated and widely integrated into production processes, enhancing process control efforts. Results so far have been impressive, with many gains made. The future looks even more promising with manufacturing professionals directly involved.

For the experienced manufacturing professional, this book provides a means of reviewing inspection and measurement concepts and gaining some new insights into how to approach them. For those new to inspection and measurement, the book offers the technology and methods for effectively planning applications. Both veterans and novices will get in-depth knowledge and understanding of inspection and measurement to help better meet the challenges they may face.

I would like to thank my professional colleagues in industry and academia who provided a forum for discussions of this important topic over the years. Such idea exchanges greatly influenced the viewpoints expressed within these pages.

My thanks also go to the Publications Development staff at SME for their assistance in making this book possible. In particular, Karen Wilhelm and Larry Binstock at SME have been exceedingly gracious in helping research the material and in editing the manuscript.

Part 1

INSPECTION

Introduction

The roles of manufacturing professionals will continue to expand. They will be more involved up-front, on cross-functional teams in planning new products. Customer-related issues will be actively dealt with by manufacturing professionals. As always, demand for product quality will continue to be the overriding issue.

INTEGRATION OF MEASUREMENT AND INSPECTION

Most would agree that new challenges present an opportunity for the manufacturing professional to design better processes than in the past. Performed up-front, the design of products can be altered, while still on paper, to aid producibility and ensure better quality. Operations can be resequenced, before tools and equipment are ordered, to achieve the best results. Inspection and measurement operations no longer need to be "standalone" and looked at as "mass inspection" in nature. Inspection and measurement operations can be integrated into the process to become the very cornerstone. Data can be gleaned from these operations to help reduce variability and fine tune the output. Key to maintaining product quality is the planning of adequate inspection and measurement strategies to provide data that regulates the manufacturing process. Handling the planning up-front gives more lead time to design and get equipment, rearrange the factory area, and train people concerned with the process.

Inspection and measurement are needed because of variability. Any process produces a product in which the features and dimensions inherently vary. Also, each time the same product is inspected or measured, different results are probable. Causes of this variability may be lack of:
- proper control of the process;
- uniformity of material or parts entering the process;

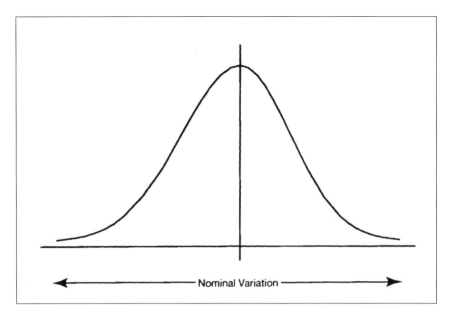

Figure 1-1. Variation with normal curve.

- control of the environment such as temperature, humidity, or vibrations; or
- control of human errors.

This variability is found throughout nature, with an exact duplicate of anything being improbable. Inspection and measurement help identify the extent of this variability. To illustrate this, Figure 1-1 shows variation likely found when measuring a specific dimension on many products. Often the data is represented by a normal curve or distribution.

Depending on results, a process may produce acceptable products, rework may be required, or an improvement in the process may be in order. This knowledge is critical to any process and will continue to be critical in the future.

In simple terms, inspection can be viewed as examining something carefully to see if it meets expectations. The uses of inspection are many, ranging from deciding whether products are okay to providing data for process improvement.

Typically, the manufacturing professional takes the lead in planning inspection points. In determining these points, decisions need to be made regarding such things as instrumentation, methods, and the necessity of records. A key evaluation is whether the inspection can be handled by sensors in the equipment planned for the process.

Inspections may start when parts or material are received, continue at in-process points, and conclude at packing and shipping. They may be handled by operators, laboratory equipment, or automatically, depending on the circumstances. All of this must be considered to come up with the best value for the process.

Like inspection, measurement is a way something can be examined, with the results expressed in numbers. Measurement is one way, among others, of performing an inspection. Lord Kelvin, a distinguished Scottish scientist, expressed the importance of measurement more than 100 years ago:

"When you can measure what you are speaking about and express it in numbers, you know something about it; and when you cannot measure it, when you cannot express it in numbers, your knowledge is of a meager and unsatisfactory kind."

This assessment by Lord Kelvin is just as valid today as it was then. One factor driving this continuing importance is that those seeking continuous improvement of processes require effective measurements to learn status and plan strategy. Typically, the manufacturing professional takes the lead up-front in evaluating where measurements are appropriate in the process. They also, at this time, evaluate such things as which type of gage or measuring equipment is most effective. Plans also may be made at this stage so that measurements can be used for controlling the process.

Some History

Inspection and measurement have been around for a long time. An Egyptian wall painting from about the 15th century B.C., in a tomb in ancient Thebes, shows an inspector assessing a stone cutter's work by measuring with a scale and plumb bob.

In fact, modern manufacturing owes its origin to inspection and measurement. The industrial revolution in the late 19th century was made possible by inspection and measurement techniques that helped achieve processes for making parts with consistent dimensions. With consistently made parts, the notion of interchangeability fostered by Eli Whitney was finally a reality. Mass production was introduced and organizations became more complex.

With this complexity, the need for separate inspection departments arose early in the 20th century. Over the first three quarters of the century, in-

spectors were increasingly relied on to assure that manufacturing operators produced good work. Because of this, many manufacturing operators
felt relieved of the responsibility for quality. Only when the inspectors
found a problem did many operators feel compelled to correct the problem. Too often, this attitude resulted in bad products being shipped to customers.

For most of the 20th century, separate inspection organizations were
common in industry. These organizations held the uncomfortable position
of being responsible for quality, but with control for producing it remain-
ing with manufacturing. Companies became very aware of the cliche that,
"you can't inspect quality into a product—you must build it in." Many
companies were more respected for their excellent service organizations
than for outgoing quality.

Time of Growth

Over this time, the inspection organizations generally grew in size. Many
adopted new names, like quality control and quality assurance, in recogni-
tion of their growth and new responsibilities. Many key advances were
made during this time, with World War II a tremendously motivating factor. New applications included such things as acceptance sampling, statistical quality control and methods, and quality standards. Later on, in the
1960s, there was a growing realization that quality could be achieved only
through the orchestrated effort of the entire company. This notion became
the key to future improvement efforts.

Seeking a Change

About 20 years ago, a change in how to deal with quality was sought by
some companies. After many years of separate inspection departments, a
change was long overdue. A different approach to quality was sorely needed
in an emerging international marketplace. In hindsight, the changes sought
at the time could be thought of as more revolutionary than evolutionary.
Probably largely driven by productivity improvements through eliminating redundancy, they had far more long-range benefits for product quality.

For starters, the standalone inspection function was transferred from
the quality organization to manufacturing. Thus, responsibility for quality
was matched with the organization that could control quality—manufacturing. Production operators checked their own work and adjusted their
machines when necessary. Figure 1-2 shows the situation where the production operator also is doing self-inspection.

Sometimes, statistical process control (SPC) was used by operators to
fine tune processes. Widespread training in quality techniques was pro-

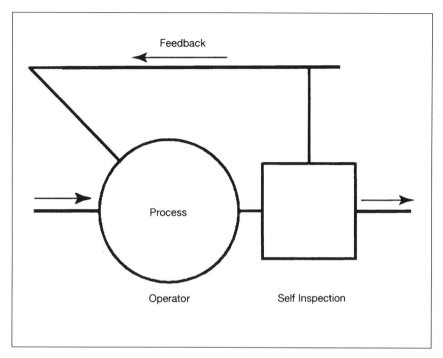

Figure 1-2. Self-inspection by operator shortens feedback loop to regulate.

vided, resulting in pride of workmanship and improved quality. Currently, the quality of many companies is competitive in the marketplace because of these changes.

Then, the quality organization was directed to act only in a consulting role to others in the company. This allowed time to develop the ideas of customer satisfaction and further motivate the entire company toward improving quality. It was a rough road, but it was increasingly clear that the way of doing business must change for a company to survive. Product development and cross-discipline teams were an immense help in making progress. Team members helped emphasize the necessity of knowing customers' needs and incorporating them into designs.

The cornerstone was laid for the evolution leading to today's quality efforts. Many changes took place: renewed customer focus, simultaneous engineering, teamwork, and empowerment. Now, inspection and measurement are no longer standalones and mass inspection is mostly gone. They are widely integrated into the role of the production operator and commonly the production equipment comprising the processes.

Inspection and Measurement

For companies that have undergone these changes, integration of inspection and measurement in manufacturing processes varies widely. Some of this variation is due to how recently the change was made. Another reason is that manufacturing personnel concerned with inspection and measurement need additional knowledge. This book is intended to help provide that knowledge.

Bibliography

Banks, Jerry, *Principles of Quality Control*. New York: John Wiley, 1989.

Grove, John W., editor, *Handbook of Industrial Metrology*. Englewood Cliffs, NJ: Prentice Hall, 1967.

Juran, J.M. and Gryna, Frank M. Jr., *Quality Planning and Analysis*, 3rd ed. New York: McGraw Hill, 1993.

Suntag, Charles, *Inspection and Inspection Management*. Milwaukee, WI: ASQC Quality Press, 1993.

Inspection and Variation

Inspection results can help determine if the product on hand is okay. If it is not, the product may be reworked or simply scrapped. There will be a cost to correct the situation, a loss in labor, and possibly material, to fix or replace the product.

Results from the inspection also can be used to help reduce variability in the process so future products are better. Also, the process could be recentered when the mean of the process is off from the nominal dimension. The mechanism of fine tuning the process is called by various names, depending on the application—statistical process control (SPC), process control, or adaptive control.

VARIATION AND ITS CAUSES

All processes make products having features and dimensions subject to variation. In fact, variation is part of everything—our environment, the things we buy, the things we sell, and everything we do. Exact duplicates of anything are unlikely.

The loss function concept, popularized in the 1980s, provides an important perspective on variation. It assumes that the specification mean or target is required by the customer and, in a larger sense, by society for optimum performance. Any variation from that target value results in an increasing loss to the customer and to society in repair costs and other expense, as illustrated in Figure 2-1. This has been verified in many studies such as those relating to transmissions produced at the Ford Batavia plant and television sets made by Sony America. The concept strongly supports a continuous effort to drive variation to the lowest possible level.

There are specific reasons or causes of variation. They can be placed in four general groups:

Common causes are those resulting from the process used. When variation is random in nature, common causes will have a consistent influence on variation if the process is not changed.

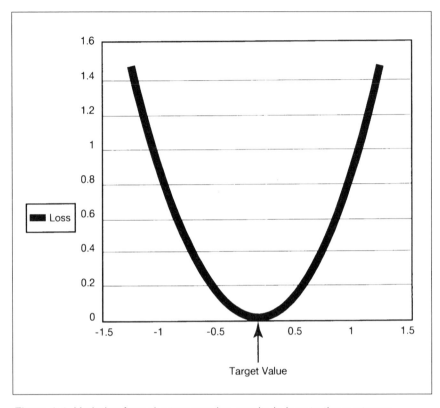

Figure 2-1. Variation from the target value results in loss to the customer.

Special causes, which can be traced to an identifiable source, produce variation that is often sporadic in nature. The variation is beyond that caused by the process. SPC is dedicated to pointing out special causes and helping remove this source of variation.

Tampering, or unnecessarily adjusting the process, causes additional variation that is not wanted. Often, this occurs because of errors in information provided by inspection to drive process control.

Structural causes are due to unique characteristics of the process being studied. For example, a machine or tool may wear out over the long term, causing unwanted variation.

Deciding which of the four types of causes produces an unwanted variation is critical to better regulating the process. Each type requires a different process control strategy. From an inspection standpoint, accurate and timely information about variation is absolutely critical to regulate the process to low levels of variation.

Why is Reducing Variation Critical?

Manufacturing processes almost always result in a variation of part dimensions and other properties, such as strength and weight, among parts produced. These, in turn, can cause a variation in the functioning of the product.

To assure that each product functions consistently, reducing variation to low levels is often critical. For example:

- In assembly operations, interchangeability of mating parts is often absolutely necessary. Without interchangeability, mass production would not be possible.
- Proper clearance must be maintained for parts not directly controlled or adjusted during assembly. If not, the product may not function as planned and its expected life may be shortened.
- When all parts of the product are assembled, they must display uniform aesthetic characteristics (surface finishes, for example).
- To enable repairs by customers, service parts must fit and function as originally planned, regardless of the sequence in which they were built.

Because of reduced variation, manufacturing losses may be less. If so, manufacturing cost also may decrease. Such a cost advantage may be decisive in the marketplace. Customers, on the average, also could experience lower operating costs, longer product lifetimes, greater reliability, and better performance.

WHAT IS A SPECIFICATION?

The objective of a specification is to identify a product's functional, durability, and appearance requirements. More formally, ISO defines a specification as a document stating product or test requirements, referring to or including drawings or patterns, and indicating the means and criteria by which conformity can be checked. Some companies include sample sizes, frequency, and acceptance criteria for certain critical quality characteristics. Besides visual examination or measurement, testing may be necessary, perhaps including testing of products during production for performance and durability.

Early in the product development cycle, features must be defined by dimensions, chemistry, or other characteristics. All this information is included in a specification. Defining the specification is often challenging, particularly when dealing with aesthetics like paint finish appearance, requiring visual inspection. Of all inspection methods, visual inspection is most widely used.

For each characteristic, a nominal or target value should be specified. For a dimension, this would be what the process is expected to make. Anticipating variation, limits above and below the nominal or target value also need to be established. These limits, called tolerances, are the outer boundaries for the envelope in which the quality characteristic must lie. This is used in inspection to judge product conformity.

Establishing realistic tolerances is also a challenge. They must be narrow enough to assure proper functioning of a product if a dimension lies in the envelope. They also must be wide enough so that the process can produce a dimension for essentially all products within the same envelope.

Many customers now require that processes have capability represented by a C_{pk} of 1.33 or higher. This means that processes must be capable of consistently producing parts with little variation. Saying it another way, processes must produce parts having so little variation that a minimum of eight standard deviations will fit within the tolerance envelope. Continuing variation reduction, despite tolerance requirements, is strongly supported by the loss function concept.

Aesthetic characteristics are often difficult to define in quantitative terms. A means of qualitatively assessing whether needs have been met should be provided. Sometimes requirements can be defined through a sketch, but often masters of the product need to be made.

Are Some Characteristics in a Specification More Important?

Quality characteristics in a specification are not equally important. A few are critical because of their effect on the product's ability to meet:

- regulatory requirements,
- safety considerations, and
- customer satisfaction perceptions.

The importance of other characteristics ranges from serious to minor. Over the years, many product designers have been reluctant to group characteristics by seriousness for many reasons. Some felt this made it easier to obtain evidence for successfully pursuing product liability lawsuits against the company. Others feared that manufacturing would not be concerned at all about those quality characteristics having only minor importance.

Now, there is a growing trend to document the "seriousness classification" of quality characteristics for a product. A typical approach for classifying the characteristics is shown in Figure 2-2. This extends to even designating a critical quality characteristic by placing a unique symbol,

---

Seriousness Classification

• Critical
   – High probability of not meeting regulations, not being safe, or failing during service causing severe customer satisfaction issues.
• Serious
   – Moderate probability of product failure causing moderate customer satisfaction issues.
• Major
   – Probability of substandard performance causing customer satisfaction issues.
• Minor
   – Probability of flawed esthetics causing customer satisfaction issues.

Figure 2-2. An approach to classifying seriousness of quality characteristics.

called a key product characteristic (KPC), on the drawing. Such information can be used to sort out the importance of the many quality characteristics presented to planners. More emphasis and greater effort can be extended to characteristics critical to the functioning of the product. This can be done without ignoring quality characteristics having a lesser consequence.

GENERAL INSPECTION CONCEPTS

The late Dr. W. Edwards Deming addressed the notion of inspection in his famous "14 points." In his third point, he exhorts industry to "cease dependence on inspection." His vision of inspection was that of a bandage, sorting good products from products that are not okay, keeping the real problem of excessive process variability on the back burner.

Dr. Deming's assessment is quite logical. Reducing variation must be placed on the front burner. When variability is reduced, quality is predictable. Processes can be more likely regulated by upstream activities and measurements—before it is too late to fix without losses occurring.

What is Inspection?

In manufacturing, "inspection" refers to the gathering of information on an output of a process, then comparing what was produced with what was needed or specified.

Inspection and Measurement

The International Organization for Standardization (ISO) defines inspection as an activity such as measuring, examining, testing, or gaging one or more characteristics of an entity and comparing the results with specification requirements to establish whether conformity was achieved for each characteristic. In its definitions, ISO also recognizes the use of inspection results for process control. It also suggests that inspection take place as close as possible to where the characteristic is made in the process. This supports integrating inspection as part of the manufacturing process.

Note that this definition considers inspection broadly as including measurement, gaging, and testing. Also, the definition requires that results be compared with specified requirements—often challenging in visual examinations.

Another way of looking at inspection is by the various tasks involved:

1. Reading and interpreting the specification.
2. Checking the quality characteristics called out in the specification by measuring, testing, gaging, and visually examining the product.
3. Comparing quality characteristics examined with the specification.
4. Judging whether the quality characteristics examined conform to the specification.
5. Passing on to the next step in the process products for which the examined quality characteristics conform with the specification.
6. Identifying and segregating, if practical, those products for which the examined quality characteristics do not conform with the specification. Usually, someone or something else, like a material review board, will be designated to decide if they will be scrapped, reworked, or used as-is.
7. Recording status in an inspection record. The record could be in a computer data base, allowing easier updating and eliminating much paperwork. ISO defines a record as a document that furnishes objective evidence of activities done or results achieved. It is proof of fulfillment of requirements for product quality or operating effectiveness of a quality system element. ISO also clarifies the meaning of objective evidence—information that can be proved based on facts obtained through observation, measurement, test, or other means.
8. Providing information on results to drive process control for regulating or controlling the process.

What Does Inspection Do?

For manufacturing, one may classify inspection as performing several functions.

Detection

It is critical that we keep a defective product from reaching a customer—for customer satisfaction and perhaps regulatory or product liability considerations. Detection is the sorting of good products from products that are not okay, before they are shipped to the customer. Products that are not okay need to be either reworked or scrapped.

Diagnosis

Without understanding the status of the process, it would be impossible to understand what corrections need to be made. Inspection furnishes information about status, which is used to drive process control so that needed corrections can be made to regulate the process.

For example, inspection can provide the dimension produced by a process. This can be compared to the specification for the dimension establishing any needed correction to bring the actual dimension produced in close agreement with the specification. Process control drives this connection.

Data Gathering

Records of manufacturing processes are needed to meet designated quality standards. One use of records is for tracing and capturing bad products already shipped. Information gathered by inspection can be used for this purpose, commonly called traceability. ISO formally defines traceability as the ability to trace the history, application, or location of an entity by means of recorded identifications. Information, useful for traceability, may contain:
- origin of materials and parts;
- product processing history; and
- distribution and location of products after delivery.

What is Expected from Inspection?

Accuracy

Information from inspection must be accurate. With inaccurate information, wrong decisions and actions are likely. To have accuracy, inspection errors must be at a minimum. Errors in inspection can distort the true variability of the process.

If errors in inspection are pronounced, information from inspection that drives process control may be distorted. With skewed information, regulation or control of the process may not be correct, making results worse instead of better.

Inspection and Measurement

Timeliness

Inspection information must be available when it's needed. If not, the benefit may be doubtful. The goal should be real-time. Anything less increases the risk that the situation has changed, making the information misleading. This properly justifies integrating inspection in the production process as close as possible to the pertinent operation. For example, production operators that check their own work can quickly adjust their processes when needed.

Flexibility

Inspection dedicated to a specific dimension or feature may not be adaptable to new needs. Rather, more flexible inspection methods may have more usefulness over the long run. Often situations change, necessitating similar inspection of different dimensions or features. Designing flexibility into inspection applications is often challenging.

Stability

Getting similar results each time from inspection requires stability. Wear and tear on instruments from repeated use should be noticeable, and corrected easily. A changing environment could affect results. This means, for example, changes in temperature, humidity, or vibrations should be monitored to assure they have little effect on inspection. Achieving stability is challenging, especially when inspections are integrated into the production process. Where humans are involved in the inspections, devices should be designed to reduce variation among operators. This may be equally challenging, requiring much ingenuity.

How is Inspection Planned?

Inspection is best planned early in the product development cycle, along with planning the process. In this way, it can be integrated in the process for better effectiveness. This technique is documented in a quality plan, which is often called a control plan. ISO defines a quality plan as a document setting out the specific quality practices, resources, and sequence of activities about a particular product, process, or contract.

Who and What Does Inspection?

In general, inspection is done by humans, or by equipment and gages, or a combination of both. Equipment, in this context, refers to both measuring and test equipment. With the gradual reduction of standalone inspection and the integration into the production process, an increasing amount of inspection is done without human intervention.

More sophisticated gages and equipment are being adopted to support the human effort, reducing inspection errors due to human frailty. Computers are widely used in human and equipment gaging applications to program procedures, help in decision making, and record results.

Done by Humans

The process of products being sorted by humans must be rethought. Besides being boring to some, the sorting process has far too many inherent errors to be effective. About one third of parts are incorrectly sorted by humans. Some industries are replacing inspection by humans with such things as in-line gaging and sensors integrated into the process. These techniques are much less prone than humans to err in sorting good from bad parts. An added benefit is that in-line automatic gaging and sensors provide data to drive process control devices. This data is necessary to limit dimensional variation.

Equipment and Gages

Equipment and gages may be integrated into processes, stand alone on the factory floor, or in environmentally-controlled company laboratories. An increase of in-process gaging, coupled with postprocess monitoring, is currently being seen. This includes closed-loop feedback to both control the process and gather data. The data gathered assesses long-term trends and spots such things as potential bearing failures or fixturing problems. Laboratory equipment is being redesigned to withstand the rigors of the factory floor.

Outside Laboratories

Outside sources, like independent laboratories, also may do inspection and certification functions for a company. In a few companies, inspection is still the responsibility of the quality department. However, there is an increasing trend for companies to assign the inspection responsibility to manufacturing. This may include inspections early during the planning and product development cycles, as well as the execution of inspection operations during production.

When should Inspection be Done?

Inspection should be done in the product development phase, as well as during production. In general, inspection should be performed at the following points.

Prototype Build. When built, usually in small quantities, these parts will be tested by product designers to confirm whether customer requirements are met. Although it is desired that prototype parts be built and checked under the same conditions as planned for production, this is seldom done. Inspection, including material and functional testing, should be conducted during the build of the prototype parts. Although much of the inspection may not be identical to that planned for production, the same intent must prevail—evaluating the conformance of quality characteristics to the requirements in the specification.

Pilot Build or Pre-launch. This is a trial run or startup, with input coming from production tooling, gaging, materials, and operators. Typically, there is much fine tuning to achieve the performance necessary to meet requirements. There will likely be additional inspection and controls until the production process is validated as meeting customer requirements. This additional effort is to contain any nonconforming parts until necessary improvements have been made. Normally, product design tests will be done using parts from this phase to validate that the production process can meet customer requirements.

Production Part Approval. Many variations in the methods are used to gain approval. Some companies, for example, require that each feature mentioned in the specification be measured to gain approval before shipping. Besides actual measurements, test results and an appearance evaluation must be done on a minimum of one piece from an extensive run. Results are compared with requirements in the specification. The part evaluated is called the master sample and must be retained for future reference. A major purpose of the large run is to find if the preliminary process performance is acceptable on critical dimensions.

Production. Inspection will continue as planned for this phase. Improvements pointed out during the pilot phase will have been implemented. Additional inspection and controls will no longer be necessary. It is critical that work continues on improving the process—and the inspections and controls associated with it.

After Production. Unfortunately, there are inspections occasionally required after the product has been shipped due to suspected nonconformance. Reaction to these can be described as knee-jerk. Planning for what to do and how to do it is usually made on the spot. Sometimes the product is returned. Other times, checks are made at the customer's facility or in the field. It is critical that findings are promptly conveyed to those concerned with the process so that corrections are made in a timely manner.

How Much Inspection Should be Done?

In today's competitive marketplace, only inspection that adds value to the process should be performed. Sometimes inspection is done because a customer requires it in the specification to help assure quality. However, inspection is often planned when the process is established, frequently by cross-discipline product development teams. Inspection might be necessary to help assure compliance with a regulation, to reduce the possibility of a safety problem or to assure customer satisfaction. Today, there is an upward trend in specifying inspection to drive process control. This may be performed by the machine operator—statistical process control (SPC)—or automatically performed in applications like adaptive control. For each quality characteristic, inspection might be handled in several ways, depending on the situation.

No Inspection

Doing no inspection is a viable alternative for most of the quality characteristics of a product. If no value is added to the process, and there is no reason to require inspection, why do it? For many companies, this stance has been adopted for the checking of parts and materials from suppliers. Rather than checking parts, they rely on the suppliers' quality systems to make parts conform to requirements.

Samples Only. Inspecting only samples means checking only part of the processed material. The approach to taking these samples and inspecting them may vary, depending on the application.

Setup. Inspecting a small group of parts before the run is started will likely provide assurance that all parts in the run will meet requirements.

Driving Process Control

SPC requires, for example, that small groups of parts be collected at intervals in the production run and inspected. This is normally done by the production operator who plots results on control charts to regulate the process.

Acceptance Sampling

A well-developed technique, acceptance sampling requires that a sample consist of parts taken randomly from a production lot. Results from the parts in the sample are assumed to represent what would be found if the entire lot was checked. When destructive testing is necessary, acceptance sampling allows testing to be done with the loss of a minimum portion of the entire lot. Although acceptance sampling is still used, many quality professionals doubt if the precision is sufficient to provide an adequate assurance of quality in today's marketplace. There is a trend to eliminate acceptance sampling with such approaches as more reliance on suppliers' quality systems and more effective process control.

100% Inspection

Inspection of some quality characteristics for all parts in a production run can be costly. However, if humans are involved, inspection errors may be intolerable. But, such inspection may be required by the customer for requirements of a critical nature. Much progress is currently being made in handling 100% inspection automatically by machines without human intervention. Such automatic applications eliminate high inspection errors due to human frailty. Such errors may be bad parts judged good or, alternatively, good parts rejected.

Where Should Inspection Occur?

Deciding the location of inspection points is done by the planners early in the product development cycle, when the manufacturing process is planned. It is usually different for each situation and, among other things, depends on the requirements to satisfy the customer. Inspection should be considered for several situations (see Figure 2-3).

If receiving inspection is used, inspection stations are located in the receiving area where parts and materials from suppliers are first seen. Based on relationships established with suppliers, receiving inspection may be considered unnecessary.

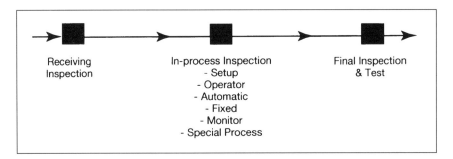

Figure 2-3. Typical locations of inspection in the process.

In-process Inspections

As Dr. Deming pointed out, inspection should not wait until a product is finished to determine problems and make corrections. In-process inspections give an early warning and a chance to correct problems before loss occurs. There are several types of in-process inspections.

Setup and first pieces inspection are of high priority to minimize the risk of nonconforming products. Properly set up, many processes run with the desired consistency for the production run. It is important that sufficient parts be selected for inspection. A large sample, maybe 25 parts, is needed to establish that the process is set at target or nominal values. Also, the large sample will assess if variation is properly controlled. This can be a good predictor on well-controlled processes. But, its prediction power is limited on processes that are chaotic in nature.

Inspection and tests by the production operator are intended to give timely feedback so that the process can be kept in fine-tune. Feedback is immediate and there is an opportunity for instant correction of problems. Sometimes the information collected is used to drive SPC efforts that aid in process control. Operators can keep their own charts on troublesome quality characteristics.

Automatic inspection and tests provide strategically placed information that can be used to drive process control devices that regulate the process. As mentioned previously, errors inherent in inspection by humans are eliminated by this equipment.

Although still in use, there is a trend to eliminate **fixed inspection stations at points during the process** in favor of inspections placed close to where the quality characteristic is determined. In this way, the information gained can be used to regulate the process in a timely manner.

Monitoring specific operations by patrol inspectors is still in use. But, the trend is to integrate inspection into the production process and use information from it for regulating the process. Patrol inspectors check various troublesome points in the process. Some view the role of these inspectors as not directly contributing to continuously improving the process.

Special process inspection is performed on products when it's not feasible to inspect certain quality characteristics. Rather, the settings of the process that make the characteristics are closely monitored. An example of this is the plating of parts where variables of the process, such as pH values, can be tracked.

Final Inspection and Tests

The final assessment is made as to whether requirements of the customer are met. Requirements pointed out in specifications are checked. Also looked at is whether all operations have been completed and processing is satisfactory. Several approaches are used. Sometimes 100% inspection is used on products that have critical applications. In other cases, sampling is used. There is a trend toward using more automated inspection equipment to better handle complicated checks.

Future Inspection Characteristics

Inspection will continue to be key for processes in the future. However, it will be different for most companies. Characteristics of future applications of inspection:

1. Less dependence on humans and more on automated inspection to reduce errors.
2. Sorting good from defective products will still be critical on those features having severe safety or customer satisfaction implications.
3. Inspection will be largely in-line, upstream, and integrated into the process to provide information to drive process controls.
4. Variation in products will be reduced and may not be a major issue with improved process controls.
5. Planning for inspection will start early in new product development cycle. It also will be done concurrently with planning the process.

WHAT IS THE NATURE OF INFORMATION FOUND DURING INSPECTION?

In quality terminology, information found by inspection can be put in several classifications.

Attribute. How many products are good is an attribute. Nine good products out of 10 products made is a measure of quality. Conversely, one not okay product out of 10 made is also a quality indicator. The latter context forms the basis for attribute process control charts, such as the P and NP chart, which were popular when go/no-go gages were widely used. Variable gaging to improve diagnostic and problem-solving capability has reduced the need for attribute process control charts.

Variable. This type of information is a measurement on a scale—for example, 0.8 inch (20 mm), 11.48 ounces (325 grams), 99° F (37° C), or 10 pounds force (45 Newtons). Variables differ from attributes in that they can be used to find how much variation there is from a desired measurement. This is useful in diagnosing and problem solving through variable gaging. Variables also can be used effectively in automatically regulating a process. Variable control charts, such as the X-bar ($\overline{X}$) and R chart, help machine operators control the variation in processes.

What are the Types of Inspection Errors?

On a general basis, the types of errors revealed during inspection are shown in Figure 2-4. Following are descriptions of these errors.

Action	Product is OK	Product is not OK
Inspection passes product	No error made	Type II error
Inspection rejects product	Type I error	No error made

Figure 2-4. Types of errors found during inspection.

Type I or Alpha Error

Type I, or alpha error, occurs when inspection concludes that the product is not acceptable, but it is really okay. At this point, a fruitless search would likely be started for the reasons that the product is not okay. Sometimes the process may be shut down causing a loss in profits from products that would otherwise be produced. Adjustments also could be incorrectly made based on this information, such as with automatic process controls. These adjustments, called tampering, could confound the process, causing a shift to the making of products that were indeed unacceptable.

Type II or Beta Error

Type II, or beta error, occurs when inspection decides that the product is okay, but it is really not. Defective products are shipped. Customer satisfaction, the key to future sales, may fall. Either rework is necessary or new products for replacement must be made. Also, extra cost in labor, and possibly material, is a financial loss.

Bibliography

Anthis, Dennis L., Hart, Robert F., and Stanula, Richard J., "The Measurement Process: Roadblock to Product Improvement." *Quality Engineering*, V3 N4, p. 461.

Aronson, Robert B., "Inspection and Quality Assurance." *Manufacturing Engineering*, August 1994, p. 184.

Hodson, William K., editor in chief, *Maynard's Industrial Engineering Handbook*, 4th ed. New York: McGraw Hill, 1992.

Joiner, Brian L. and Gaudard, Marie A., "Variation, Management and W. Edwards Deming." *Quality Progress*, December 1990, p. 29.

Inspection and Measurement

Juran, J.M., editor, *Quality Control Handbook,* 4th ed. New York: McGraw Hill, 1988.

Juran, J.M. and Gryna, Frank M. Jr., *Quality Planning and Analysis,* 3rd ed. New York: McGraw Hill, 1993.

Liggett, John V., *Dimensional Variation and Management Handbook: A Guide for Quality, Design and Manufacturing Engineers.* Englewood Cliffs, NJ: Prentice Hall, 1993.

"Quality Management and Quality Assurance–Vocabulary." International Organization for Standardization (ISO/DIS) 8402, 1992.

Salvendy, Gavriel, editor, *Handbook of Industrial Engineering.* New York: John Wiley, 1992.

Suntag, Charles, *Inspection and Inspection Management.* Milwaukee, WI: ASQC Quality Press, 1993.

Winchell, William, *Continuous Quality Improvement: A Manufacturing Professional's Guide.* Dearborn, MI: Society of Manufacturing Engineers, 1991.

Planning for Inspection
Using the Control Plan

Planning for inspection is largely based on the experience and judgment of those involved. In general terms, planning for inspection of a variable includes the following steps:

- Choosing the variable to be measured, for example, a dimension.
- Choosing a unit by which to measure the variable. Such a unit may be a millimeter.
- Designating a sensor to measure the variable. In a manual application, it may be a micrometer. A transducer, such as a probe, may be used for automatic inspection.
- Selecting a goal for the variable. Often, this is the specification from the product drawing.
- Deciding the measurement frequency.
- Making the measurement using the sensor.
- Finding the difference between the measurement and the goal for the variable.
- Reporting the difference found. This may be used for judging whether the product is acceptable or for regulating the process.

Checking attributes, such as the aesthetics of a product, follows generally the same procedure as used for variables. However, working with attributes is much more challenging and often is based exclusively on the judgment of a person. The specification is typically defined via a master, an acceptable product selected from early production run. The inspector judges whether the product selected for examination meets the standards of the master. There has been some success in using such things as optical sensors to automate what would otherwise be a manual subjective process.

CONTROL PLAN

A control plan documents the characteristics of the products, as well as the processes involved, in helping assure a quality product. The plan also includes the inspections, process controls, and tests required.

Inspections could measure a dimension of a product or a process characteristic. On the other hand, inspection could be judging, in qualitative terms, a product's aesthetic qualities. Results from inspections may be used to judge if what was made is okay.

Data from inspection also could drive process control. In this way, it is used to adjust processes so that product quality is maintained.

Tests could relate to checking the characteristics of the material processed. They could also provide information as to the performance and durability of the products being manufactured. Results could be compared with specifications to decide if products meet desired requirements.

Development and Manufacturing Require Different Control Plans

Usually, there are specific control plans for a product while it is developed, and other plans for that same product during its manufacturing phase. The limited build of prototype parts for testing during the product development cycle may require quite a different control plan than that for production.

The few units built for prototypes do not justify a large up-front investment before production. Yet the need for information on prototypes is usually greater than that for production units. Consequently, extensive manual inspections and testing of a general nature will normally dominate the control plan for prototype parts. Measuring devices for achieving this are usually simple and readily available.

A larger investment for production is necessary to achieve the quality objectives for the higher volume repetitive operations. Usually, the means for achieving this are specifically designed for the application. A control plan for production often has fewer inspections. These inspections, which may be specialized in nature, are increasingly being done automatically. Manual inspections using simple measuring devices are not widely used. Also, control plans for production often have process control provisions and accelerated specialized testing.

At startup of production, a special control plan may be appropriate. Inspection may be intensified at this time to catch an anticipated higher rate of rejects. Also, it may be necessary to provide redundant checks to verify results from specialized applications that are being debugged.

Each Type of Process Requires a Different Emphasis

Processes can be categorized as to the source of variables having the greatest influence on quality. For each source, a different emphasis may be appropriate when the specific control plan is developed.

The following examples are not exhaustive. Yet they do illustrate some elements that may be included in control plans. In reality, a control plan usually addresses a process influenced by more than one of the aforementioned conditions. Consequently, each control plan is usually custom designed to account for specific conditions.

Equipment Setup

This type of process is highly capable and very stable but must be set up properly for the production run. In essence, the machine involved is highly repetitive and is not significantly influenced by external variables during the production run. The control plan may include:

- verification that machine settings correspond to that specified for setup;
- inspection of first piece(s) made after verification of machine settings; and
- lot control, if the part is of a critical nature, between setups.

Machine Settings

For this, the dimensions of the variables involved in the process may change during the production run and, because of this change, the process may not meet quality requirements. To control this, the changing dimensions of the variables are measured and the process is adjusted to stay within acceptable limits. The control plan may specify inspection of significant variables to drive the process control effort. Sometimes, automatic inspection may be appropriate. Also, self-adjusting devices may be specified for the significant process parameters that change quickly. The control plan also may include variable statistical process control charts on the significant process and product dimensions that change more slowly.

Variation Among Fixtures

Variation among fixtures causes parts to be rejected for not meeting dimensional or other quality requirements. Other causes may be chips or other debris deposited on the locating surfaces of the fixtures. For this, the control plan may emphasize inspection to check on the cleaning of fixture-locating surfaces, adjustment of fixtures to reduce variation, and the creation of variable statistical process control charts for the significant product dimensions that may be affected.

Design of the Tool

The shape or design of the tool is the largest contributor to whether or not the part meets quality requirements. Often, the tool's shape directly forms the shape in the part. An example may be the pitch diameter of a thread. For this process, the control plan may specify checking the dimensions of the tool before use, and inspecting first piece(s) made.

Wear or Breakage of Tools

In certain processes, tool wear and breakage affect whether a part meets quality requirements. For example, a part may have missing holes. Activities that may be appropriate for the control plan are automatic inspection for such things as missing holes, automatic inspection of significant variables to drive process control, self-adjusting devices on the variables that change because of wear, and variable statistical process control charts on the critical process variables.

Scheduled Maintenance

Many machines require periodic maintenance such as cleaning, lubricating, calibration, and part replacement. Processes are typically quite sensitive to whether this maintenance is performed. For these processes, the control plan may include scheduled maintenance programs, warning devices to signal when scheduled maintenance is not done, and inspection of significant product or process variables to verify that maintenance was done correctly.

Operator Differences

Adequate training of the operator and the operator's ability to control the process has a great bearing on assuring product quality. The control plan may include training requirements for the operator and inspector (for example, training on how to make statistical process control charts).

Material or Components Variation

Variation in the composition or size of materials and components entering a process affects the outcome. For some materials, shelf life is also very important to product quality. Control plans for this situation may include:

- surveillance of suppliers;
- lot control for shelf life;
- laboratory reports for composition;
- tests for physical properties; and
- inspection to make statistical process control charts on the significant process and product variables.

Variation in the Environment

Changes in humidity and temperature may affect a process, causing products to not meet quality requirements. Also, temperature changes may result in significant variation in the measurement of dimensions. For these reasons, controlling the environment for some processes is critical. Control plans for these types of processes may specify automatic control of appropriate environmental variables. An example may be installing heating ventilation or air conditioning and a thermostat to control the temperature of the environment surrounding the process.

How Much Inspection is Needed?

Valuable insight into the need for inspection can be found by examining customer requirements, the design failure mode and effects analysis (DFMEA), and the process failure mode and effects analysis (PFMEA). How much inspection can vary from none to 100% of the product for each of the various characteristics and features. There are several considerations in making the decision.

Customer requirements. Certain features may be considered critical and require 100% inspection. Also, the first parts produced may require more extensive inspection than parts made after first piece approval.

Stage in the development/production cycle. The stage in the development/production cycle determines how much inspection is needed. For example, prototype parts may be examined more extensively than parts in production.

History of similar products. Quality experience, such as surveys, warranty records, and customer complaints on similar products may help predict the inspections needed.

History of suppliers. Quality assessments of suppliers may help decide how much inspection is needed.

Criticality. Certain characteristics and features may be critical to product performance. Also, it may be that certain characteristics and features are critical to the successful completion of subsequent production or service operations. It may be best to inspect the characteristics and features to avoid problems later on. Normally, this need for inspection would be pointed out by DFMEA and PFMEA.

Process capability. Processes that are properly controlled would need a minimum of inspection. Such would probably be the case for those processes using automatic inspection that drives process control devices. Also, control charts used by the production operator may preclude the need for inspection. However, processes that are likely to be sporadic, like during production startup, may need intensified inspection.

Measurement capability. Redundant inspections may not be necessary if there is confidence in the measurement devices, such as the use of automatic gages.

Nature of process. A stable process is dependent primarily on how well it was initially set up. Here, inspection of only the first and last few pieces in the run may be necessary.

Product consistency. If the products being made are expected to have little variation, inspecting large sample sizes is not necessary. Also, it may not be necessary to repeat inspections frequently.

Data available on process. If data is readily available on the important process variables that influence product quality, then other inspections may not be necessary.

Process Control

For process control applications, inspection is needed to acquire data to regulate the process. Inspection may be done manually by the operator for plotting statistical process control charts. Also, it may be done automatically and used directly for driving process control devices within the manufacturing equipment. Automatic inspection for driving process control mechanisms is increasingly designed into production processes. Whether manual or automatic, the input to process control is from inspection. For example, the actual dimension being produced is compared to the specification, and process control makes any needed adjustments to bring the actual dimension produced in close alignment with the specification.

Operator Inspection for Process Control

This type of inspection allows the operator responsible for process results to control the process. Data is plotted on process control charts by the operator. When the process goes out of control due to special causes, the operator makes corrections. In this way, the process is regulated. Sometimes basic changes need to be made, for example, to tooling or gaging. The operator then needs to involve others in more extensive work.

Automatic Inspection for Process Control

Automatic inspection provides data to electromechanical devices that regulate and reduce variation in the process. Although more costly initially, it enables processes to provide products with little variation. In one application, a critical dimension is measured on 100% of the parts machined. Through automatic feedback to the machine, tooling for that dimension is immediately adjusted to reduce variation. Sensors for measurements in assembly inspection operations are used for:

- traditional gaging, such as touch sensors;
- color monitoring, such as infrared sensing optics;
- monitoring for force, pressure, and shock; and
- acoustic monitoring.

The type of sensor required is often determined by the ingenuity of those specifying it. For instance, a sensor was used by one company to detect steel inserts in an aluminum wheel. Instead of using a complex and expensive sensing mechanism, a simple proximity switch was specified and worked just fine.

Automatic inspection has many advantages over manual inspection. For process control devices, it is absolutely necessary. Precise and accurate data is needed instantaneously after measurements are made. This is not possible with manual inspection. Besides the slowness of manual inspection, human error is intolerable for providing data to process control devices.

Acceptance

To judge whether a product is acceptable, inspection is needed. Automatic inspection for acceptance is growing in use, but much is still done manually.

Automatic Inspection for Acceptance

Many test and inspection stations are built as part of the manufacturing equipment. For the most part, they do not drive process control devices. Nevertheless, they do provide value to a manufacturing process in that problems are identified early. Also, the frequency and location of the problems are tracked for fixing later by human effort, and the nonconforming product is isolated and directed to, for example, a rework station for prompt correction. In this way, tearing down a complex assembly to make a repair can be avoided. Inspection and repairs can be done when the feature is accessible.

Manual Inspection for Acceptance

For inspection by human effort, error often distorts the data collected. This error, inherent in manual operations, occurs because of:

- incorrect interpretation of specifications, particularly when aesthetics are judged;
- variation in the inspection method used, whether by the same operator or different operators;

- inattention; or
- conscious errors.

Efforts have been made to reduce human error in manual inspections with little success. Studies have shown that this inherent failure is due to things that cannot be fixed by redesigning the inspection process. Findings of these studies show that manual inspection accuracy increases as nonconformities and product complexity decreases. Accuracy is also related to repetition of the inspection task. This relationship only exists up to a maximum of six repetitions. Time required to do the inspection task was found to have no bearing on human error. It appears from these studies that accuracy of manual inspection cannot be improved by redesigning the manual inspection method. Therefore, they provide strong support for using automatic inspection devices if adequate justification exists.

Planning the Plan

The best time for preparing control plans is during product development, when many companies use multidiscipline teams engaged in what is often called simultaneous, or concurrent, engineering. Members of these teams are specialists from manufacturing, product engineering, quality, and other sources such as suppliers, and even customers. Conceptual thinking, product design, production planning, and problem solving occur simultaneously when the teams meet. By working together toward a common goal, "walls" that impede progress are removed and more innovative solutions are typically developed.

This approach shortens the time spent developing a product because activities are done in parallel. For example, manufacturing processes can be designed as the product is developed. Maybe even more important, both products and processes can be carefully examined from the different perspectives of the various team members. Changes can be made while designs are still on paper to make both the product designs and processes better. In this way, better quality products are made for less cost with a shorter time to reach the market. Often disruptive engineering changes, made after designs are committed, are also reduced.

For many companies, distinct control plans are usually required at various stages in the product development cycle. Early in the cycle, a control plan for prototype parts is needed. Later, separate control plans are needed for the startup of production and then the production phase. They may be quite different in nature and composition.

With the product development cycle, each plan can be designed at the point of use. By involving a team, diverse viewpoints are considered from the various disciplines. This approach also allows knowledge learned in the execution of earlier plans to be used in developing later plans. Such advantages will likely result in better control plans.

Planning for the product design, production process, and inspection is best done simultaneously. Quality is really dependent on the product design and production processes. Studies show that product design is typically responsible for about 80% of quality and production processes for about 20%. It is widely understood that you cannot inspect quality in a product—it must be designed and built in.

But, inspection is needed to help assure quality by driving process control and providing data for acceptance decisions. To make this work, the design of the product and process, and the desired support from inspection, must be closely coordinated. All three must operate in concert.

The approach for design for manufacturing and assembly (DFMA) encourages planning for error-proof assembly ("poka-yoke"). This may take some innovative approaches in designing the product and the production processes. If successful, inspection would probably not be needed in the control plan to check if the assembly was acceptable.

Surveys have shown that using DFMA to support development efforts has been extremely successful regarding quality. On the average, a 68% improvement in quality and reliability was accompanied by a 37% reduction in part costs, a 62% drop in assembly times, and a 57% reduction in manufacturing cycle times.

Consistent location of the part must be maintained for the product design, equipment, tools, and measuring devices to assure quality. Many companies rely on geometric dimensioning and tolerancing (GD&T) concepts. GD&T is used to define the product so that it meets the intent of the design. It is also used to provide direction for the design of production processes and measuring devices. GD&T symbols are used on a product drawing to show allowable variation in dimensions for various features. Key to doing this is the specification of datums or locating surfaces and what sequence they must recognize. The use of datums is shown in Figure 3-1. Using the same datums or locating surfaces consistently as specified when setting up the equipment, tools, and measuring devices is essential.

Unfortunately, this sometimes does not happen and inconsistent parts are made. If proper datums, for example, are not maintained in measuring devices, good parts may likely be wrongly rejected. One study showed that 99% of parts passed an inspection gage when following GD&T specifications correctly. But when improper datums and positioning sequence

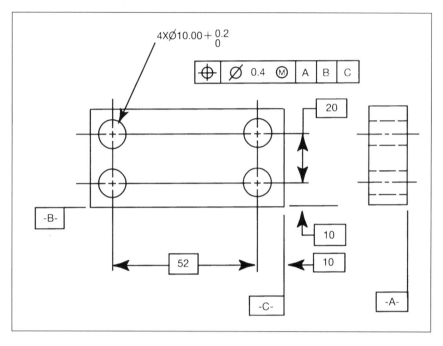

Figure 3-1. Use of datums in GD&T. Datums are labeled A, B, and C.

were used, only about 20% of the same parts passed the gage. By planning simultaneously for the product design, production process, and inspection, inconsistent application of GD&T concepts is less likely to happen.

Planning for inspection *after* the design of the product and process is completed could result in faulty applications. Unfortunately, this is a common problem when inspection is done as an afterthought. In one case, access for inspection and test depended on how a product was designed. There was not enough room in an assembly fixture to assemble a product and test it in-line too. If the problem had been caught early in the product development cycle, the solution would have been a simple design change in the product. But, it was too late to do this and a costly off-line inspection and test station was needed for the duration of production.

Typical Method of Doing a Control Plan

Typically, making a control plan involves a design failure mode and effects analysis (DFMEA), process flow chart, and process failure mode and effects analysis (PFMEA) as shown in Figure 3-2. Detailed operator instructions are a by-product of the control plan.

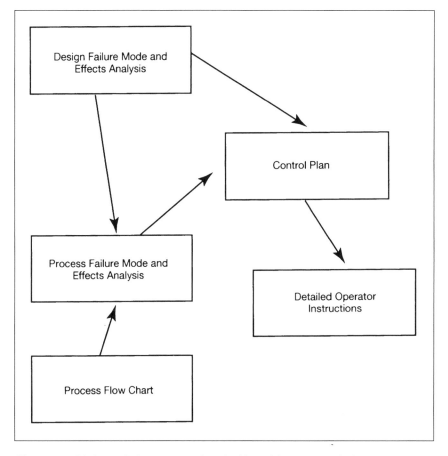

Figure 3-2. Major techniques associated with making a control plan.

Design Failure Mode and Effects Analysis

Developing a control plan typically starts with input from the product design. Most often this is by a DFMEA. Other methods may be used such as a quality function deployment study or fault tree analysis. However, most companies use the DFMEA because it is easier and has proven effective over several decades. It was first used formally in the aerospace industry in the mid-1960s and was widely adopted by the automotive industry during the 1980s.

Information from the DFMEA can help those preparing the control plan to identify manufacturing and assembly requirements. It can also act as a catalyst for ideas about optimizing the strong relationship among product design, process design, and materials.

Preparing the DFMEA is done by manufacturing professionals with design responsibility. It is done early by the multifunctional product development team. A manufacturing professional, working for a major supplier having no design responsibility, could also be asked to help in this effort.

In simplistic terms, DFMEA is a summary of what may go wrong in the product design based on the experiences and concerns of the members of the product development team. This summary is used to help eliminate causes of failures or reduce their effects. Action taken by the product development team because of the DFMEA includes changing the product design and revising the validation procedures to assure design intent. The potential failures identified by the DFMEA could relate to such things as:

- safety;
- customer satisfaction;
- performance;
- reliability;
- environmental concerns;
- regulations and standards;
- durability and life expectancy; and
- serviceability.

One could view the DFMEA as a disciplined analytical tool that assesses the probability of failure modes of a product design and its associated causes. A potential failure mode is the way in which a product design could fail to meet the design intent. It may occur, but not necessarily. Examples of a potential failure mode could include leaking or fracturing. The cause of failure could be viewed as a design weakness. Causes could include wrong material or insufficient lubrication. Also, a DFMEA helps identify the effect of such failure on the product's function as seen by the customer. Examples of effects are noise, impaired operation, or bad appearance.

The DFMEA is also a means of helping reduce the risk of potential failure modes in designs and their associated causes. Product design changes and modification of validation procedures are commonly done as part of the analysis. The validation procedures use laboratory tests and other engineering means to establish that the product design meets the design intent.

The DFMEA must be done early in the product development cycle. It is here, while product designs are still on paper and least costly to change, that necessary modifications can be made to reduce the risk of failures. A

DFMEA should be considered a living document, frequently updated as the needs and expectations of customers change or are identified.

For those doing the control plan, the column labeled Risk Priority Number (RPN) is critical. This is the measure of risk in the design for each specific concern. The RPNs for the various concerns should be ordered by rank to set priorities for fixing. Numbers could range from 1 to 1,000. Concerns having higher RPNs should receive the greatest priority for corrective action. For all practical purposes, sufficient resources may not be available to fix those concerns with lower RPNs. However, each specific action taken is a matter of judgment after carefully considering the risk involved.

RPN is determined for each concern by multiplying the values in the columns labeled "Sev" (severity), "Occ" (occurrence), and "Det" (detection). A widely used DFMEA scheme in the automotive industry, available from AIAG, defines these terms as follows.

- "Sev" (severity) measures the impact of potential failure modes or design weakness on an internal or external customer. It is important to stress that severity only concerns the effect on the customer. Also important is that a reduction in severity can only be achieved by a product design change.
- "Occ" (occurrence) is the likelihood that a specific product design weakness or cause of a failure mode will occur during the life of the product while in the hands of the customer. This rating only can be reduced by a product design change that removes or controls the cause of the failure mode.
- "Det" (detection) evaluates the ability of the current product validation procedures to identify a design weakness before a product is released for production. This rating only can be reduced by changing the product design validation procedures. To fully understand the risk involved for a particular concern, it is helpful to know how these numbers are established (see Table 3-1).

The individual values for "Sev," "Occ," and "Det" are important to forming a judgment regarding the advisability of corrective action. For example, many believe a high severity number justifies corrective action despite the RPN number.

It is important to understand the definitions of severity, occurrence and detection used in the specific DFMEA. For example, the definitions used widely in the automotive industry, and in this chapter, limit corrective action to product design or validation procedure changes. This type of

Table 3-1. DFMEA Rankings Used in the Automotive Industry

"Sev" or Severity	*Basic Concern*	
9-10	Safety exposure or noncompliance with government regulation	
7-8	High degree of customer dissatisfaction	
4-6	Moderate degree of customer dissatisfaction	
2-3	Slight customer annoyance	
1	Unnoticeable by most customers	
"Occ" or Occurrence	*Basic Concern*	*Failure Rate*
9-10	High failure	1:2–1:3
7-8	Repeated failure	1:8–1:20
4-6	Occasional failure	1:80–1:2,000
2-3	Few failures	1:15,000–1:150,000
1	Unlikely failure	1:1,500,000
"Det" or Detection	*Basic Concern with Design Validation*	
10	No detection of design weakness	
9	Probably will not detect design weakness	
7-8	Unlikely to detect design weakness	
5-6	May detect design weakness	
3-4	Good chance of detecting design weakness	
1-2	Almost certain to detect design weakness	

DFMEA does not depend on process controls or inspections to overcome weaknesses of the product design. But, this DFMEA and the product design it is based upon does take the intended manufacturing process into consideration on such things as limitations on surface finish, access for tooling, and inherent process capability. Corrective actions may include the clear calling out of design intent in product drawings, specifications, and other manufacturing documents. Drawings or other documents also may include a classification of characteristics for certain features and dimensions requiring additional process controls and inspections.

Process Flow Chart

Typically, the next major task in doing a control plan is developing the process flow chart. This chart represents the anticipated manufacturing sequence and the relationship of steps in that sequence. It is important that each process or operation be labeled with a unique number, name, and a concise description. This information will be used when making the PFMEA and the control plan.

Often special process and product characteristics requested by the customer are put on the flow chart by relevant operations. A concise description of how the function of a product is affected by various design failure modes may be included.

Figure 3-3 is an example of a process flow chart. A living document that should be prepared early in the product development cycle, it is modified frequently as better ways of making a product are found by members of the multifunctional team. By using the flow chart, the entire process can be focused upon and analyzed.

Process Failure Mode and Effects Analysis

A typical process failure mode and effects analysis (PFMEA) is similar in concept to a DFMEA except the PFMEA is used to analyze the production process. Like the DFMEA, it does this by using a disciplined review to define potential process problems. If problems are not resolved, actions can be taken so that they are monitored and controlled.

Making the PFMEA occurs early in the product development cycle before the control plan is drafted. Normally, all members of the product development team take part, making the PFMEA a catalyst that promotes the interchange of ideas among team members. It is modified frequently as better ways of doing things are identified by team members.

The PFMEA is an analytical tool to reduce the likelihood of process failure modes and their associated causes. In many ways, it is a summary of what could go wrong in the process based on the experiences and concerns of product development team members. These problems can then be dealt with and resolved. The PFMEA lists:

- potential manufacturing process failure modes;
- effects on the customer by the potential process failures;
- causes of the potential process failures;
- process variables that must be focused on to reduce the risk of process failures; and
- ranking of the potential process failure modes providing a rationale for corrective actions.

Those preparing the PFMEA assume that the product design will meet the design intent. Thus, potential failures due to product design are not included. However, these potential failures are addressed in the DFMEA completed earlier in the product development cycle.

Inspection and Measurement

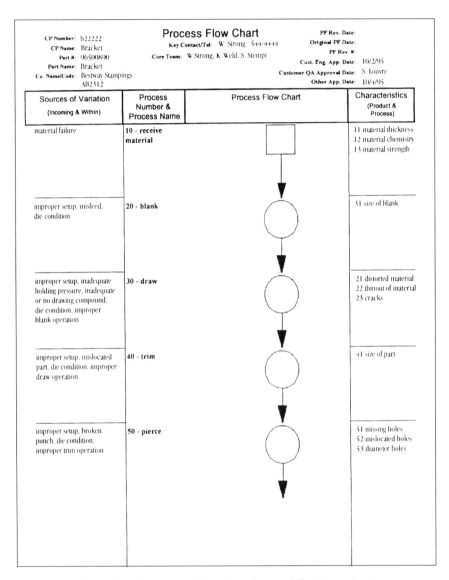

Figure 3-3. Example of a process flow chart (Integral Solutions, Inc.).

In preparing the PFMEA, changes in the product design are not relied upon to resolve process weaknesses. Nevertheless, those preparing the PFMEA should fully consider product design characteristics in planning and evaluating the ability of the process to meet customer needs. In a few cases, the PFMEA may reveal that the process cannot make the proposed product design under any feasible circumstances. An iterative discussion may then be necessary to determine an alternative process.

The following describes what should be entered in the major columns of a PFMEA form (Figure 3-4).

Process Name. The purpose of the operation or process being analyzed is concisely expressed in this column. A product design may have several operations or processes. Sample entries may include such things as drilling, welding, or assembly.

Potential Failure Mode. This column contains a concise description of how a production operation or process may not meet the expectations or product design requirements of the customer. More than one potential failure mode could be listed for each production operation or process. In other words, it is the possible types of nonconformance due to that specific operation or process. An assumption is made that any incoming material is correct. Examples of column entries could be such things as binding, cracked, dirty, or deformed.

Potential Effect(s) of Failure. How the customer is affected by a potential failure mode is expressed in this column. As with a DFMEA, the customer could be viewed as either internal or external. If you're not shipping directly to the ultimate user, it is important to consider both types of customers. In this way, the effect on subsequent operations is considered. Some examples of entries are cannot fasten, does not fit, or does not match. Effects of product performance on the ultimate user should also be stated. Examples of effects include erratic operation, poor appearance, or rough.

Potential Cause(s)/Mechanism(s) of Failure. This describes succinctly how the potential failure could happen. It is important that causes be described in terms that can be fixed or controlled. Ambiguous phrases should be avoided. More than one cause could be assigned to each potential failure mode. Some examples of causes are inadequate lubrication, part missing, or inaccurate gaging.

Current Process Controls. A brief description of existing controls in the process to prevent or detect each potential failure mode is made here. This could be such things as inspections, process control devices, SPC charts, or fixture error proofing. Location of these controls may be at the operation or in stations after the operation.

RPN. As with the DFMEA, the column labeled "RPN," or Risk Priority Number, is critical. This measures the risk in the production process for each specific concern. RPNs for the various concerns should be ordered by

Potential Failure Mode and Effects Analysis (Process FMEA)

Co. Name/Code: Bestway Stampings — AB2312
Process Resp.: W. Strong
BP Change Date: 9/2/95
Other Areas Affected: n/a
CP Number: b22222
CP Name: Bracket
Department #: 446
Dept. Name: Press Room
Process: 9/15/95

Affected Plant: Seabreeze, NY
Veh Line/Model Year: Skybird 1997
Machine #:
Part #: 96500890
Part Name: Bracket
Process Number: 10
Process Name: Receive Material

Customer QA Approval Date: S. Louvre
Cust. Eng. App. Date: 10/2/95
Plant Approval Date: 9/22/95
Other App. Date: 10/4/95
Original FMEA Date:
Revised FMEA Date:
FMEA Rev. #:
Key Contact/Tel: W. Strong, 544-4444
Core Team: W. Strong, K. Weld, S. Stempi

Characteristic ID/Name	Potential Failure Mode	Potential Effect(s) of Failure	Sev	Class	Potential Cause(s)/ Mechanism(s) of Failure	Occ	Current Process Controls	Det	RPN	Recommended Actions	Responsibility & Target Completion Date	Action Results — Actions Taken	Sev	Occ	Det	RPN
11-material thickness	Oversize	Can't complete process	7		Supplier manufacturing	3	Failure Cause: Supplier process control	4	84	ISO 9000 registration	Purchasing 12/1/95	ISO 9000	7	3	2	42
	Undersize	Breakage while customer using causing noise	7		Supplier manufacturing	3	Failure Cause: Supplier process control	4	84	ISO 9000 registration	Purchasing 12/1/95	ISO 9000	7	3	2	42
12-material chemistry	Incorrect	Breakage causing noise/can't complete operations	7		Supplier manufacturing	2	Failure Cause: Supplier process control	4	56	ISO 9000 registration	Purchasing 12/1/95	ISO 9000	7	2	2	28
13-material strength	Incorrect	Breakage causing noise/can't complete operations	7		Supplier manufacturing	2	Failure Cause: Supplier process control	4	56	ISO 9000 registration	Purchasing 12/1/95	ISO 9000	7	2	2	28

Department #:446
Dept. Name: Press Room

Process Number: 20
Process Name: blank

Characteristic ID/Name	Potential Failure Mode	Potential Effect(s) of Failure	Sev	Class	Potential Cause(s)/ Mechanism(s) of Failure	Occ	Current Process Controls	Det	RPN	Recommended Actions	Responsibility & Target Completion Date	Action Results — Actions Taken	Sev	Occ	Det	RPN
31-size of blank	Undersize	Can't complete operations	5		Misfeed	5	Failure Cause: Feed mechanism	6	150	Operator check	Manufacturing 12/1/95	Operator check	5	5	4	100

Figure 3-4. Process failure mode and effects analysis (PFMEA) (Integral Solutions, Inc.).

rank to set priorities for fixing. Numbers could range from 1 to 1,000. Concerns having higher RPNs should receive the greatest priority for resolution. RPN is determined for each concern by multiplying together the values in the columns labeled "Sev," "Occ," and "Det." It is important to understand the definitions of severity, occurrence and detection used in the specific PFMEA. To fully understand the risk involved for a particular concern, it is also helpful to know how these specific values were established.

- **Sev.** "Sev" (severity) measures the impact of potential failure modes or process weakness on an internal or external customer. It is important to stress that severity only concerns the effect on the customer. Also important is that a reduction in severity can only be achieved by a product design change. The number used for severity should correlate to that used for the DFMEA (see Table 3-2).

- **Occ.** "Occ" (occurrence) (see Table 3-3) is the likelihood that a specific weakness or failure mode will occur during the execution of the process. Available failure detection measures are not considered with this evaluation. This rating can be reduced by process design changes that remove or control the cause of the process failure mode. A product design change may also help in certain cases.

- **Det.** "Det" (detection) (Table 3-4) evaluates the ability of current process controls to identify a process failure mode before a product leaves the production location. This rating can be reduced by changing the process. However, making detection better is often very costly. Reducing occurrences or preventing the problem from happening is logically the first priority. Product design revisions may be appropriate sometimes to help in detection.

- **Class.** "Class" or classification designates any characteristic that may require more inspection or process control due to its importance to the overall design intent. Often this is called a "Seriousness Classification" of the characteristics discussed in Chapter 2. They may be called out by the customer or members of the product development team. The class designates the importance of the characteristic to customer satisfaction, safety issues, regulatory compliance, general product design intent, or manufacturing process stability.

Recommended Action(s). This is a concise description of the corrective action taken to lower RPN.

Responsibility and Target Completion Date. A list of who is going to do the corrective action and when it will be done.

Action Results. This records the completion of corrective action and the revised RPN because of the corrective action.

**Table 3-2. PFMEA Severity Rating Scheme Used
in the Automotive Industry**

"Sev" or Severity	*Basic Concern*
9-10	Safety exposure or noncompliance with government regulation
7-8	High degree of customer dissatisfaction
4-6	Moderate degree of customer dissatisfaction
2-3	Slight customer annoyance
1	Unnoticeable by most customers

**Table 3-3. PFMEA Occurrence Rating Scheme Used
in the Automotive Industry**

"Occ" or Occurrence	*Basic Concern w/Process*	*Failure Rate of Process*
9-10	Very high failure	1:2–1:3
7-8	High failure	1:8–1:20
4-6	Moderate failure	1:80–1:2,000
3	Low failure	1:15,000
2	Very low failure	1:150,000
1	Unlikely failure	1:1,500,000

Table 3-4. Detection Rating Scheme Used in the Automotive Industry

"Det" or Detection	*Basic Concern with Controls to Detect before Leaving Area*
10	No detection of defects
9	Probably will not detect defects
7-8	Unlikely to detect defects
5-6	May detect defects
3-4	Good chance of detecting defects
1-2	Almost certain to detect defect

Control Plan Summary

The purpose of a control plan is to assure the production of quality products that meet customer needs. It provides a structured approach to control methods that add value to a production process. The control plan documents the quality system designed to reduce variation in both the product and process. It is a living document, constantly updated to reflect better ways of doing things for the lifetime of the product.

A control plan contains the actions required at various stages of the process. This includes actions to be taken for receiving, in-process, and outgoing products. It also includes other broad-based actions that are done periodically. During production, the control plan specifies the process monitoring and control schemes that reduce variation of both process and product characteristics. A characteristic can be viewed as a feature, property, or dimension of a process or product for which data can be collected. The data can be either variable or attribute in nature.

The objective of a control plan (Figure 3-5) is to help assure that all outputs of the manufacturing process are in control. Detailed operator instructions that carry out the intent of the control plan can be prepared using the control plan as a general guide. It is important to understand that a control plan is not intended to provide these instructions directly to the operator.

Development and maintenance of a control plan helps improve customer satisfaction, quality, and communications. Customer satisfaction is improved because the control plan focuses limited resources on controlling process and product characteristics important to the customer. Quality is improved because the control plan identifies the sources of variation in process and product characteristics. Controls can be adopted to reduce the troublesome sources of variation, allowing a more consistent product and reducing waste. Communication is improved since members from multiple disciplines participate in making a control plan and take part in the steps leading to it. Thus, both needs and changes are expressed without delay to those directly concerned with producing the product.

The control plan content depends greatly on the experiences and concerns of those on the multidiscipline team. Those on the team will typically find the following quite useful in preparing the control plan:

- process flow chart;
- DFMEA;
- PFMEA;
- special characteristics ranked by a scheme such as a seriousness classification (could be done either by the customer or product development team);
- history of similar parts and any future plans;
- results of design reviews; and
- tools such as design of experiments or Ishikawa cause-and-effects diagrams to analyze processes for things like sources of variation.

The following describes what should be entered in the major columns of the control plan.

Inspection and Measurement

Prototype ☐ Pre-launch ☐ Production ☐					**Control Plan**						

CP Number: b22222 CP Name: Bracket
Part #: 96500890
Part Name: Bracket
BP Change Date: 9/2/95
Co. Name/Code: Bestway Stampings   AB2312
Plant Approval Date: 9/22/95

Key Contact/Tel: W. Strong, 544-4444
Core Team: W. Strong, K. Weld, S. Stempi

Original Control Plan Date:
Revised Control Plan Date:
Control Plan Revision #:
Cust. Eng. App. Date: 10/2/95
Customer QA Approval Date: S. Louvre
Other App. Date: 10/4/95

Part/Process Number	Process Name/Operation Desc.	Machine, Device, Jig, Tools for Mfg.	No.	Product	Process	Special Char. Class.	Product/Process Specification/Tolerance	Evaluation/Measurement Technique	Sample Size	Freq.	Control Method	Reaction Plan
10	Receive material		11	Material thickness			0.040 .002	Supplier			Supplier manufacturing: Supplier process control	Undersize: Check incoming material. Oversize: Check incoming material
			12	Material chemistry			1012 ±spec	Supplier			Supplier manufacturing: Supplier process control	Incorrect: Check incoming material
			13	Material strength			70.000 ±5%	Supplier			Supplier manufacturing: Supplier process control	Incorrect: Check incoming material
20	Blank		31	Size of blank			±	L1234 Length gage	5	Hourly	Misfeed: Feed mechanism	Undersize: Stop. Fix.
30	Draw		21	Distorted material			±	v10000 Visual check	5	Hourly	Improper setup: Setup wrong	Distorted: Stop. Do setup.
			22	Thinout of material			±	v10000 Visual check	5	Hourly	Improper setup: Setup wrong	Thin material: Stop. Do setup.
			23	Cracks			±	v10000 Visual check	5	Hourly	Improper setup: Setup wrong	Cracks: Stop. Do setup.
40	Trim		41	Size of part			±	f563 Form gage 2SC-4400	5	Hourly	Mislocated part: None	Undersized: Stop. Train operator.
50	Pierce		51	Missing holes				v10000 Visual check			Punch broken: Die missing punch	Holes missing: Stop. Fix die and setup.
			52	Mislocated holes			2.000 +.005	12314 Gage			Improper setup: Setup wrong	Mislocated holes: Stop. Do setup.

Figure 3-5. One page of a control plan (Integral Solutions, Inc.).

Part/Process Number. This item number correlates to that used on the process flow chart.

Process Name/Operation Description. This name or description correlates to that used on the process flow chart.

Machine Device, Jig, Tools for Manufacturing. The processing equipment (machine, jig, fixture, or other tools) is listed for each operation.

Characteristics. Lists the cross reference number from such things as the process flow diagram, PFMEA, or prints.

Product. Special product characteristics are listed. These may have been called out by the customer or the product development team. Other product characteristics also may be listed for which process controls are normally provided.

Process. Lists the process characteristics or variables that have a cause-and-effect relationship with the product characteristics. There may be more than one process characteristic for each product characteristic.

Special Characteristic Class. Lists any special classification for a characteristic such as the "seriousness classification." This is furnished by a customer or decided by the product development team.

Product/Process/Specification/Tolerance. Contains the specification and tolerance for each product or process characteristic.

Evaluation/Measurement Technique. Lists the system used to measure product or process characteristics. It could include gages, fixtures, tools, or test devices.

Sample Size/Frequency. Contains details of any sampling plan that is used.

Control Method. Lists concisely the control method used for the product or process characteristic. It may be by such activities as SPC, inspection, mistake proofing, or sampling plans.

Reaction Plan. This is the expected corrective action to be taken if the process starts to go out of control.

Bibliography

Advanced Product Quality Planning and Control Plan. Southfield, MI: Automotive Industries Action Group (AIAG), 1994.

Aronson, Robert B., "Inspection and Quality Assurance." *Manufacturing Engineering*, August 1994, p. 184.

Ashley, Steven, "Cutting Costs and Time with DFMA." *Mechanical Engineering*, March 1995, p. 74.

Besterfield, Dale H., et al, *Total Quality Management*. Englewood Cliffs, NJ: Prentice Hall, 1995.

Boubekri, Nourredine and Schneider, Morris H., "Inspection and Testing Methods for Manufacturing Processes." *Quality Engineering*, Vol. 3 No. 4, p. 491.

Hardt, David E., Fenn, Ralph C., and Bakkestuen, Robert S., "Feedback in the Forming Press." *Metal Forming*, March 1995, p. 45.

Hodson, William K., editor in chief, *Maynard's Industrial Engineering Handbook, 4th ed.* New York: McGraw Hill, 1992.

IMPACT-Integrated Manufacturing and Control Tool, Software Version 2.2. Royal Oak, MI: Integrated Solutions, Inc., 1995.

Juran, J.M., editor, *Quality Control Handbook, 4th ed.* New York: McGraw Hill, 1988.

Juran, J.M. and Gryna, Frank M., Jr., *Quality Planning and Analysis, 3rd ed.* New York: McGraw Hill, 1993.

Juran, J.M., *Juran on Leadership for Quality*. New York: The Free Press, 1989.

Liggett, John V., *Dimensional Variation and Management Handbook: A Guide for Quality, Design and Manufacturing Engineers*. Englewood Cliffs, NJ: Prentice Hall, 1993.

Measurement Systems Analysis. Southfield, MI: Automotive Industries Action Group (AIAG), 1990.

Noaker, Paula M., "Sensible Sensing for Assembly." *Manufacturing Engineering*, September 1992, p. 49.

Potential Failure Mode and Effects Analysis in Design (Design FMEA) and *Potential Failure Mode and Effects Analysis in Manufacturing and Assembly Processes (Process FMEA Reference Manual)*. Warrendale, PA: Society of Automotive Engineers, SAE J1739, 1994.

Production Part Approval Process. Southfield, MI: Automotive Industries Action Group (AIAG), 1993.

Quality Management and Quality Assurance - Vocabulary. International Organization for Standardization (ISO/DIS) 8402, 1992.

Quality Systems—Model for Quality Assurance in Design, Development, Production, Installation and Servicing. Milwaukee, WI: American Society for Quality Control ANSI/ASQC Q9001-1994.

Salvendy, Gavriel, editor, *Handbook of Industrial Engineering.* New York: John Wiley, 1992.

Video: *Linking Measurement to the Machine — Manufacturing Insights.* Dearborn, MI: Society of Manufacturing Engineers, 1995.

Wearring, Collin and Karl, Dennis P., "The Importance of Following GD&T Specifications." *Quality Progress,* February 1995, p. 95.

Standards for Inspection

The number of standards used in the world is huge, nearly uncountable; among them are standards concerned with quality, products, measurements, testing, and weights. Some standards are generic and many others are for specific industrial sectors or individual companies. Standards of various countries may deal with the same subject, but vary widely in approach. Although growing in number, there are few truly international standards that are common throughout the world.

SOURCES OF STANDARDS

About 11,000 standards are cataloged for just one standards writing organization in the United States—the American National Standards Institute (ANSI). Other major writers of standards in the United States include professional organizations like the American Society of Mechanical Engineers (ASME), American Society for Quality Control (ASQC), American Society for Testing and Materials (ASTM), Institute of Electrical and Electronic Engineers (IEEE), and the Society of Automotive Engineers (SAE). Many standards promulgated by professional and technical organizations are published by ANSI. They receive final approval by ANSI and carry both the designation of ANSI and the originating organization.

The United States government is also a major source of standards. Federal agencies active in providing standards include the Department of Defense (DoD), Environmental Protection Agency (EPA), and the National Highway Safety and Transportation Agency (NHSTA).

International Standards

Countries in every part of the world have organizations active in standards writing. In response to the worldwide interest in standards, the International Organization for Standardization (ISO) was formed. Its purpose is to develop common standards to help the interchange of goods and services throughout the world. The members of ISO are various standards bodies representing more than 90 countries. ANSI is the U.S. representative. International standards are issued on needed areas after attempts are made to harmonize the interests of each country. These international standards are then adopted by interested countries as a national standard.

QUALITY SYSTEM STANDARDS

This section refers to standards for a generic quality system which is not particular to any specific industry but, in general, applies to all industries.

A quality system deals with the organizational structure, responsibilities, procedures, processes, and resources needed to assure quality of the deliverables provided. The quality system exists both inside and outside a company. The internal system is concerned with the processes in the company, like inspection. Externally, the quality system deals with the processes of suppliers. All processes are linked together in the quality system, providing an orchestrated approach to assuring quality.

The first generic quality system standards were developed in the U.S. by the federal government after the Korean War. These were called mandatory standards in that compliance with them was required to fulfill government contracts. Examples were the broad "Military Specification Quality Program Requirements" (MIL-Q-9858) and the narrower "Military Specification Inspection System Requirements" (MIL-I-45208). Both were developed by the DoD and are still used today. These standards have equivalent requirements for inspection.

The first voluntary generic quality system standard was written by ASQC and published by ANSI in 1971. It was called "Specification of General Requirements for a Quality Program," ANSI Standard Z1.8-1971. Inspection requirements were included. "Voluntary" means that the use of the standard by an organization was not mandatory. Having no teeth to force compliance, the standard was thought of as advisory in nature and used as a guideline, if at all. Although well thought out and conceptually sound, not much formal use was made of this standard or the several versions that succeeded it by organizations in the U.S. Rather, companies drafted their own requirements, like General Motor's "Targets for Excellence" and Ford's "Q-101 Quality System Standard." Audits of suppliers were conducted by each company, using these criteria.

The first widespread use of voluntary generic quality system standards was with the ISO 9000 series of standards. These international standards were first issued in 1987 and the teeth that forced compliance was registration.

ISO 9000 Registration

Registration is performed by third parties called registrars. Companies accept the determinations of these registrars usually without conducting audits themselves. Registration of suppliers assures companies that the supplier has an adequate quality system that is periodically monitored.

Registration involves the assessment of compliance with the ISO 9000 standards by third parties. Registration is provided to the supplier if, in simplistic terms:

- the documented quality system meets the intent of the appropriate standard; and
- the supplier is doing what it promises in its documented quality system.

Maintaining the registration is contingent upon the supplier successfully completing periodic audits, usually twice a year, after passing the initial assessment.

It is perceived that, by meeting the ISO 9000 standards as verified by registration, a company will be more successful when seeking profitable international business. Fostering this perception is the multinational nature of potential suppliers and customers. Great distances do not permit supplier audits by each individual customer. ISO registration makes good business sense, avoiding much cost and unwise use of resources.

Registration Advantages

Business improvement through ISO 9000 registration has become reality and more companies are seeking registration. Many organizations around the world have established quality systems based on the ISO 9000 standards. By mid-1994, there were more than 70,000 organizations worldwide registered in compliance. This is projected by ISO to climb to more than 200,000 organizations in the next four years.

In addition, both contractual and industrial sector support is growing. In the U.S., DoD and NASA have recently issued "Guidance on the Application of ISO 9000/ASQC Q90 Series Quality System Standards" as MIL-HDBK-9000/NASA-HDBK-9000. Instead of other military standards, ISO 9000 standards and registration can now be used to meet the requirements of government contracts. The automotive sector has also recently issued "Quality System Requirements" as the QS 9000 standard. This uses the ISO 9000 standards as a base, with other sector and company requirements added. Registration for QS 9000 also includes ISO 9000 registration. Other industry sectors in the U.S., such as chemical and furniture, also demand that suppliers have ISO 9000 registration.

Other advantages derived by a supplier from ISO 9000 registration are:

1. The number of customers that accept ISO 9000 registration as supplying adequate confidence in a company's quality system is growing. This means that audits and surveys by these customers probably will not be needed.
2. With fewer audits, there is less disruption of normal operations and less use of resources for audits and surveys.
3. With one standard, instead of differing requirements from each customer, compliance is easier to achieve and maintain.
4. Companies registered have a positive quality image widely known to other potential customers.

ISO 9000 History

ISO first published a set of five international generic quality system standards in 1987 after the approval of most member countries. These were promptly adopted in the U.S. as national standards and known as the Q90 standards. More than 60 other countries also adopted these as national standards.

Prior to these standards, many years were spent attempting to harmonize the interests of the various member countries. Existing mandatory and voluntary quality system standards from different countries were used as building blocks for the ISO standards.

In drafting these standards, state-of-the-art requirements for a quality system were not sought. Rather, only minimum requirements were included so that adequate quality could be delivered. In this way, a broad base of companies could be in compliance with the ISO standards across the world. Registration of a company signifies an adequate quality system, not necessarily an exceptional one.

Although ISO standards are felt to be superior to MIL-STD-9858, they are not felt to be as stringent as specific industry sector standards, like QS 9000, the requirements for the Malcolm Baldrige Award, or the Deming Prize. Requirements of the ISO standards are viewed as minimum good business practices for a supplier in any industry sector.

The more stringent requirements place more emphasis on such things as customer satisfaction and continuous improvement. ISO 9000 standards would be a good first step, but they would not provide a quality system that could compete for a Malcolm Baldrige Award or Deming Prize.

ISO updated the generic international quality system standards in 1994 after more than four years of harmonization effort among the member countries. During this time, all members had ample opportunities to propose, comment, and vote on suggested changes.

The Q9000 Quality System Standards

In the U.S., the Q9000 quality system standards were promptly adopted as national standards and are known as:

1. "Quality Management and Quality Assurance Standards–Guidelines for Selection and Use." Milwaukee, WI: American Society for Quality Control ANSI/ASQC Q9000-1-1994.

2. "Quality Systems–Model for Quality Assurance in Design, Development, Production, Installation, and Servicing." Milwaukee, WI: American Society for Quality Control ANSI/ASQC Q9001-1994.

3. "Quality Systems–Model for Quality Assurance in Production, Installation, and Servicing." Milwaukee, WI: American Society for Quality Control ANSI/ASQC Q9002-1994.

4. "Quality Systems–Model for Quality Assurance in Final Inspection and Test." Milwaukee, WI: American Society for Quality Control ANSI/ASQC Q9003-1994.

5. "Quality Management and Quality System Elements–Guidelines." Milwaukee, WI: American Society for Quality Control ANSI/ASQC Q9004-1-1994.

The updated standards, commonly referred to as the Q9000 series, have maintained continuity with the 1987 standards. It is also worthwhile to mention that the updates did not change the basic approach and structure of the standards. But, the changes did improve usability, especially for third party registration. Much improvement also was made in both format and the ease of interpretation. Figure 4-1 shows the relationship among the five national standards, commonly called the Q9000 series, adopted by the U.S.

Road Map

ANSI/ASQC Q9000-1-1994 can be viewed as a road map for using the other standards. Besides interpreting fundamental quality concepts, it contains key definitions and has guidelines for how to specify what standard or portion of a standard may be best for any given situation. Choices for registration purposes are limited to three possibilities.

1. ANSI/ASQC Q9001-1994 is specified for an organization when it has both product design and manufacturing responsibility. This stan-

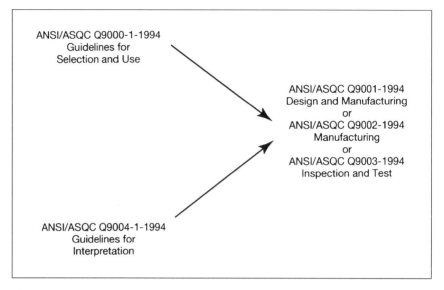

ANSI/ASQC Q9000-1-1994
Guidelines for
Selection and Use

ANSI/ASQC Q9001-1994
Design and Manufacturing
or
ANSI/ASQC Q9002-1994
Manufacturing
or
ANSI/ASQC Q9003-1994
Inspection and Test

ANSI/ASQC Q9004-1-1994
Guidelines for
Interpretation

Figure 4-1. Relationship among the five national standards commonly called the Q9000 series.

dard has the most comprehensive requirements for a quality system in the Q9000 series. It is used when a supplier must provide assurance of conformance to all requirements involved in designing, developing, producing, installing, and servicing. Inspection is a major portion of the standard.

2. ANSI/ASQC Q9002-1994 is specified for a manufacturing organization that farms out product design. This standard has the next most comprehensive requirements for a quality system in the Q9000 series. It is used when a supplier must provide assurance of conformance to all requirements involved in producing, installing, and servicing. A major portion includes inspection processes. For registration purposes, this standard is the most popular so far.

3. ANSI/ASQC Q9003-1994 is specified for an organization when it has, in general, only final inspection and test responsibility and does no product design or manufacturing. This specification has the least requirements for a quality system in the Q9000 series. It is used when a supplier must solely provide assurance of conformance to requirements at final inspection and test. So far, this specification is used the least for registration purposes.

ANSI/ASQC Q9004-1-1994 provides further guidance and interpretation useful for a supplier in implementing a quality system.

Figure 4-2 lists the contents of ANSI/ASQC Q9001-1994 and the 20 quality system elements specified as requirements. The less comprehensive ANSI/ASQC Q9002-1994 and ANSI/ASQC Q9003-1994 list identical quality system elements. However, the description of quality elements in the more limited standards may eliminate or reduce the need for rigorous requirements, as appropriate. For example, there is no need for the design requirements in ANSI/ASQC Q9002-1994.

1. Scope
2. Normative reference
3. Definitions
4. Quality system requirements
    4.1   Management responsibility
    4.2   Quality system
    4.3   Contract review
    4.4   Design control
    4.5   Document and data control
    4.6   Purchasing
    4.7   Control of customer-supplied product
    4.8   Product identification and traceability
    4.9   Process control
    4.10 Inspection and testing
    4.11 Control of inspection, measuring, and test equipment
    4.12 Inspection and test status
    4.13 Control of nonconforming product
    4.14 Corrective and preventive action
    4.15 Handling, storage, packaging, preservation, and delivery
    4.16 Control of quality records
    4.17 Internal quality audits
    4.18 Training
    4.19 Servicing
    4.20 Statistical techniques

Figure 4-2. contents of "Quality systems-model for quality assurance in design, development, production, installation, and servicing."

The Pyramid

The Q9000 standards require that the quality system of a company be documented. How it is documented is shown in Figure 4-3. The linking of the various documentation could be viewed as a pyramid.

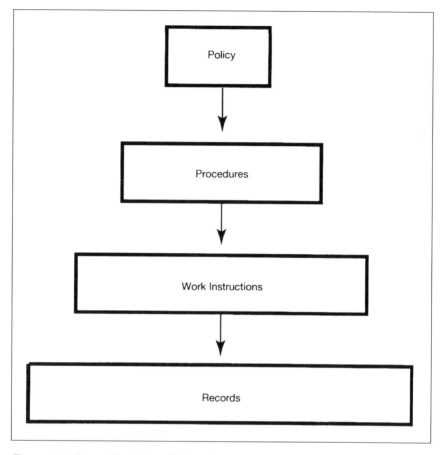

Figure 4-3. Pyramid relationship forming the basis of documentation in Q9000 standards.

The top block, the *quality policy manual*, is drafted first. This, in simplistic terms, conveys what will be done by the quality system, and why it must be done. It should be clear, concise, precise, and practical. Rarely does this document exceed 20 to 30 pages.

Second from the top are the *quality procedures*, a series of documents defining methods for carrying out the quality policy. The procedures list includes: who will do specific tasks, when they will be done, and where actions taken are documented.

Work instructions, the third layer, explain in detail how specific jobs should be done. A good test of a work instruction is if the job can be done by an employee by simply referring to the instruction. Often, work instructions are detailed steps confined within a department that may concern operating a machine or using a gage. They are frequently drafted by those doing the specific job, which helps assure buy-in. Work instructions are the most detailed documentation.

Records are the bottom layer and foundation of the documentation pyramid. They verify that the policies, procedures, and work instructions were followed and the intent of the quality system was implemented. Records provide traceability of actions taken regarding specific deliverables. Review of records can validate that a quality system is working effectively. Also, records can be used to identify corrective action required.

Writing the documents for a quality system based on ISO is probably the easiest part of the job. Making the policy, procedures, and work instructions work effectively is probably the most difficult. The key to making this happen is buy-in of those involved in doing the job.

Standards, by their nature, lag behind state-of-the-art technology or practice. But, the next revision in the generic international quality systems standards is expected to incorporate many advanced provisions of the Malcolm Baldrige Award and the Deming Prize. Quality is a moving target, and what is state-of-the-art now will likely be common practice in five years.

Inspection Provisions

Among the many activities addressed by ANSI/ASQC 9001-1994 is inspection. Each activity is composed of processes and accompanying requirements. There are several requirements for the inspection activity. The responsibilities and authority for inspection processes must be clear. This, along with organizational freedom, should allow reaching any assigned objectives.

Resources must be provided by management to achieve objectives associated with the inspection activities, including competent personnel. This also includes having the proper equipment, tools, gages, instrumentation, and software.

Quality Plans

As part of quality planning, quality plans should be prepared and updated as needed. Plans are needed particularly on new products, and should include such things as process steps and responsibility. They also cover such requirements as specifications; cycle time; aesthetics; where, when, and how to inspect, test, and audit; specific procedures and work instructions; and how and when the plan is to be updated.

The methods outlined in the last chapter for preparing a control plan could be used for the quality plan. For all practical purposes, a control plan and a quality plan are essentially the same thing.

Document and Data Control

Documentation and records are critical to a quality system. The quality manual, quality plans, procedures, and work instructions are considered documentation and define a company's quality system. Other documentation includes drawings and specifications. Ways must be provided for controlling these documents, allowing updates as needed only by authorized parties.

Records are the outputs of a quality system. They may include such things as inspection reports, test data, survey and audit reports, and material review reports. Records show if products meet requirements and verify that the quality system was functioning properly. Like documentation, records also must be controlled to assure integrity. After a prescribed time, they can be destroyed.

Verification of Purchased Products

Manufacturers must have a clear understanding of how the quality of their products will be assured. Options include:
1. Rely on the subcontractor's quality system.
2. Require inspection/test data from subcontractor.
3. Require 100% inspection/test by subcontractor.
4. Lot acceptance sampling or sorting by subcontractors before shipping to manufacturers.
5. Registration to ISO 9000 standard by subcontractor.

Materials received must be properly controlled. Areas must be provided in which materials can be quarantined while awaiting disposition.

If recalls may be an issue, material should be uniquely identified from the time of receipt to when the product is shipped. Records should identify location of the materials at each stage in the build.

Process Control

Inspection provides a link to process control efforts, usually verifying that processes can make products within specification. This verification should be broad-based and include material, equipment, computer systems and software, procedures, and people. After verification, monitoring, or inspecting and controlling, the processes should relate directly to finished product specifications or other internal requirements. If product characteristics cannot be easily measured, process variables should be monitored

or inspected and controlled. Objectives of the process measurements may be to control and improve:
- accuracy and variability of equipment;
- operator skill, capability, and knowledge;
- accuracy and variability of inspection results; and
- environment in which the process resides.

Inspection and Testing

Inspection and testing should be documented. It may be substantiated in the procedures, control plan, or work instructions. Necessary records also should be specified in these documents.

Receiving inspection and testing may not always be necessary. Many companies in the U.S. choose to rely on the subcontractor's quality system to provide assurance of incoming quality.

Inspection and tests should be at appropriate points in the process to verify conformance with requirements. In general, these points should be close to where the characteristics are first formed. Verifications are done:
- at setup;
- by the operator;
- automatically;
- at fixed stations in the process; or
- by patrolling inspectors.

The product should not be released unless steps have been taken so that it can be located later for any needed repairs or substitutions.

To verify that the finished product meets requirements, final inspection and testing should be done. Some view this step as a confirmation of the in-process checks. The method of verification could be acceptance inspections and tests such as lot sampling, continuous sampling, or 100% inspection, or it could be an audit of the representative products. No product should be released for shipping until necessary actions, such as repairs, have been made.

Inspection and Test Status

All products should be clearly identified as to whether they are in conformance with requirements. Nonconforming products should be segregated until a disposition can be reached. Choices are:
- rework to meet requirements;
- accept by concession;
- regrade for other applications; and
- scrap.

Corrective action takes the form of eliminating the cause of non-conformities and other problems.

Internal quality audits should be done periodically to verify effectiveness of the quality system, including whether quality activities bring the desired results. Those involved in the audit should be independent of the people in the audited area to avoid any bias. Management should take timely action to correct deficiencies, and confirmation of these corrections should be a priority in follow-up audits.

A plan should be prepared for training all people concerned with activities affecting quality. Strategy and time frame for the training should be recorded. People performing specific jobs should be qualified based on training received, as well as education and experience.

Statistical Techniques

Procedures should be established to select and apply statistical methods to such things as process capability studies, determining quality levels in sampling plans, data analysis, performance assessment, nonconformity analysis, and process improvement. Statistical methods that could be considered include design of experiments, regression analysis, significance tests, SPC charts, and statistical sampling.

It is important to note that the Q9000 standards tell *what* must be done and leave *how* and *when* to the supplier. A registrar checks whether all requirements in the standard have been complied with and if a company is doing what it promised.

Bibliography

Besterfield, Dale H., et al, *Total Quality Management.* Englewood Cliffs, NJ: Prentice Hall, 1995.

Clements, Richard Barrett, "Understanding the 1994 Revisions to ISO 9000." Grand Rapids, MI: National ISO 9000 Support Group, 1994.

Durand, Ian G., Marquardt, Donald W., Peach, Robert W., and Pyle, James C., "Updating the ISO 9000 Quality Standards: Responding to Market Place Needs." Milwaukee, WI: *Quality Progress*, July 1993, p. 23.

"Guidance on the Application of ISO 9000/ASQC Q90 Series Quality System Standards." Department of Defense/NASA: MIL-HDBK-9000/NASA-HDBK-9000, 1994.

Juran, J.M., editor, *Quality Control Handbook, 4th ed.* New York: McGraw Hill, 1988.

"Military Specification Inspection System Requirements." Department of Defense MIL-I-45208A, 1981.

"Military Specification Quality Program Requirements." Department of Defense MIL-Q-9858A, 1962.

Peach, Robert W., editor, *The ISO 9000 Handbook, 2nd ed.* Fairfax, VA: CEEM Informational Services, 1994.

"Quality Management and Quality System Elements–Guidelines." Milwaukee, WI: American Society for Quality Control ANSI/ASQC Q9004-1-1994.

"Quality Management and Quality Assurance Standards–Guidelines for Selection and Use." Milwaukee, WI: American Society for Quality Control ANSI/ASQC Q9000-1-1994.

"Quality Systems–Model for Quality Assurance in Design, Development, Production, Installation, and Servicing." Milwaukee, WI: American Society for Quality Control ANSI/ASQC Q9001-1994.

"Quality Systems–Model for Quality Assurance in Production, Installation, and Servicing." Milwaukee, WI: American Society for Quality Control ANSI/ASQC Q9002-1994.

"Quality Systems–Model for Quality Assurance in Final Inspection and Test." Milwaukee, WI: American Society for Quality Control ANSI/ASQC Q9003-1994.

"Quality System Requirements." Southfield, MI: Automotive Industry Action Group (AIAG) QS 9000, 1994.

Salvendy, Gavriel, editor, *Handbook of Industrial Engineering*. New York: John Wiley, 1992.

Suntag, Charles, *Inspection and Inspection Management*. Milwaukee, WI: ASQC Quality Press, 1993.

Tsiakals, Joseph J., "Revision of the ISO 9000 Standards." Milwaukee, WI: ASQC 48th Annual Quality Congress Proceedings, 1994.

Wadsworth, Harrison M., "Standards for Tools and Techniques." Milwaukee, WI: ASQC 48th Annual Quality Congress Proceedings, 1994.

Winchell, William, *Continuous Quality Improvement: A Manufacturing Professional's Guide*. Dearborn, MI: Society of Manufacturing Engineers, 1991.

Guidelines for Inspection

HELP FROM THE MALCOLM BALDRIGE AWARD

Created in 1987 by Public Law 100-107, the Malcolm Baldrige Quality Award provides insight into what is needed to go beyond a basic quality system like ISO 9000. It concerns the continuous improvement of both the quality of deliverables and performance of the business. An enhanced system for quality, motivated by the Malcolm Baldrige criteria, would very likely provide a stronger base for inspection processes.

The Seven Categories

The award has two results-oriented focuses. One is the delivery of ever-increasing value to customers. The other is the improvement of overall company performance and goals. The award's guidelines form the foundation of an effective improvement effort. Seven categories (Figure 5-1) describe the broad nature of what is expected. Some items are considered more critical than others. Different points or weights are given to each category. However, in reality, an effective effort cannot exist without all seven categories pulling together.

Customer Focus and Satisfaction

One of two categories that share the largest weight, customer focus and satisfaction, is a full one-fourth of the total points needed for the award. This category reflects how well the company understands the voices of its customers. Much of this understanding comes from measuring results and trends. Items for this category are:
1. Knowledge of the customers and the markets.
2. Management of relations with customers.
3. Determination of customer satisfaction.
4. Customer satisfaction results.
5. Comparison of customer satisfaction with competition.

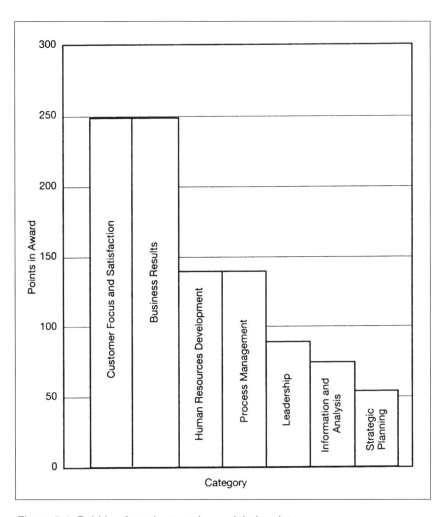

Figure 5-1. Baldrige Award categories and their points.

Business Results

Sharing the largest weight with customer focus and satisfaction is business results. Together, the two categories account for one-half the total points. The criteria of business results reflect a company's performance in the marketplace and also among competitors. Items comprising this category are:

1. Improvement results of product and service quality as reflected by key indicators.

2. Improvement results of operational and financial conditions as reflected by key indicators.
3. Improvement results of supplier performance as reflected by key indicators.

Human Resource Development and Management

This category is next in importance comprising 14% of the total award points. It concerns how a company manages its people to reach full potential. Also, it pertains to involving everyone in quality efforts. Items in this category concern:

1. Human resource planning and evaluation to support strategic and business goals.
2. Work systems that encourage achieving high performance.
3. Education, training, and employee development that address company plans.
4. Work environment and climate that encourage the well-being and satisfaction of employees.

Process Management

Process management shares the same emphasis as human resource development and management. Together, they form close to one-third of the total points. Process management concerns how key processes in a company are designed to meet customer and operational requirements. The items comprising this category are:

1. Design and introduction of new products and services.
2. Management of product and service production and delivery processes.
3. Management of support services.
4. Management of supplier performance requirements.

Leadership

Leadership relates to the culture of the company and accounts for 9% of the award. The ability of top management to create values and then spread them through the company is a major part of this category. Items in this category are:

1. Leadership of senior executives in developing and maintaining an environment for excellence in performance.
2. Leadership system and organization for customer focus and performance expectations, including the existence of these values within the company.
3. Public responsibility and corporate citizenship.

Information and Analysis

This category considers the effectiveness and management of information that helps assure excellence in customer-driven performance and success in the marketplace. It carries 7.5% of the total points. Items in this category are:

1. Management of information and data.
2. Competitive comparisons and benchmarking.
3. Analysis and use of data in reviews at the company.

Strategic Planning

The strategic planning category relates to how the company develops and carries out both short- and long-range plans. Business drivers, as outcomes of these plans, should be deployed to strengthen a customer-focused competitive position and the financial performance of the company. Strategic planning accounts for 5.5% of the total award.

More information is available on the Malcolm Baldrige Quality Award from the National Institute of Standards and Technology, Gaithersburg, MD 20899-0001.

HELP FROM INDUSTRY-SPECIFIC DOCUMENTS

QS 9000 Quality System Requirements was adopted by Chrysler, Ford, General Motors, and the major domestic truck producers as an automotive industrial sector standard in 1994. It was published by the Automotive Industries Action Group (AIAG). This agreement resulted from several decades of seeking to harmonize individual company standards.

The goal of QS 9000 is to provide the automotive industry with a fundamental quality system that pursues continuous improvement, prevents defects, and reduces variation and waste. The companies involved in drafting QS 9000 are committed to working with suppliers to assure customer satisfaction through these actions.

Now being phased in, QS 9000 will soon be mandatory for first tier internal and external suppliers to the automotive industry. A first tier supplier sells directly to an original equipment manufacturer (OEM). Whether suppliers further down on the sales chain must comply is up to the customers. Third party registration for QS 9000 may be required by customers to eliminate separate audits. For this, ISO 9000 registration must be sought first. QS 9000 contains additional requirements beyond those specified for ISO 9000.

Automotive manufacturing uses many processes, including the production of parts made from steel, cast iron, plastics, aluminum, and copper.

This industry is also a major factor in the use of computer and sensor technology. Naming a process or technology that has not been adopted in some form in this industry is difficult. Although QS 9000 is specific to one industry, it is generic in an industry that is extremely broad in scope.

Like the Malcolm Baldrige Quality Award, QS 9000 provides insight into what is needed to go beyond a basic quality system. It too is concerned with continuous improvement of both the deliverables and business performance. An enhanced system for quality, such as one motivated by the QS 9000 criteria, would very likely provide a stronger base for inspection processes.

Unlike the Malcolm Baldrige Quality Award, QS 9000 provides differences from ISO 9000 that are quite specific and perhaps easier to adopt. Many differences concern inspection processes, which involve both the planning and delivery of products.

Three Sections

QS 9000 is made up of three sections as shown in Figure 5-2. The individual sections are:

1. Section I - includes all the ISO 9001 Section 4 requirements. Specific automotive requirements and clarifications also are included that amplify and strengthen the ISO 9001 requirements. A supplier that does not have design responsibility is exempted from meeting the design requirements of ISO 9001 Section 4.

2. Section II - contains generic requirements specific to the automotive industry that strengthens QS 9000. In addition to any requirements in Section I. These QS 9000 requirements include the process for production part approval, actions for continuous improvement, and methods for improving manufacturing capabilities.

3. Section III - has unique requirements for specific companies, beyond any requirements in Sections I and II. For example, Ford has requirements concerning control item parts, critical characteristics, setup verification, control item fasteners, heat treating, and others.

QS 9000 has differences from ISO 9000 standards that are worthwhile to consider for strengthening the inspection planning and delivery processes. Following are examples of the amplifications and modifications in Section I.

Responsibility and Authority

A multidisciplinary approach for decision making must be used by suppliers. Typical functions include engineering, manufacturing, quality, service, and marketing. Suppliers must have a formal, documented business

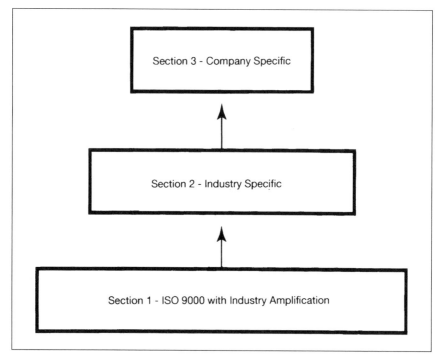

Figure 5-2. The three sections of QS 9000.

plan, including both short- and long-term goals and plans based, for example, on competitive products and benchmarking. This is intended to encourage strategic planning.

Trends in quality, operational performance, and current quality levels must be documented by the supplier. Trends should be compared with business plans and appropriate action should be taken. Trends in customer satisfaction and key indicators of customer satisfaction must be documented by the supplier and reviewed by senior management.

Quality Plan

The quality plan, called the control plan, covers prototype, prelaunch, and production. To prepare and review this plan, the supplier must use a multidisciplinary team, following certain designated AIAG references. One is the *Advanced Product Quality Planning and Control Plan Reference Manual*, published by AIAG, which includes guidelines for preparing control plans for prototype, prelaunch, and production. Checklists are provided to ensure the intent of the planning is met. A multidiscipline team environment for product and process development is relied upon.

The other document, *Production Part Approval Process*, covers generic requirements for approval of all production and service commodities, including bulk materials.

Control plans are living documents and must be updated when:
* the product is changed;
* the processes are changed;
* the processes become unstable; or
* the processes cannot meet requirements.

During preparation of the control plan, the planning team must complete the calling out of special characteristics. Symbols used by each customer for special characteristics are contained in Section III of QS 9000.

Team preparation and review of failure mode and effects analyses are required. Actions must be taken by the team to reduce potential failure modes with high risk priority numbers. Techniques are detailed in the *Potential Failure Mode and Effects Analysis Manual*, published by AIAG. This manual introduces potential failure mode and effects analysis and provides general guidance on how to apply the technique. It includes information on design, system, or process failure mode and effects analyses.

Process Control

A process must be provided that assures compliance with all applicable government safety and environmental regulations. This includes handling, recycling, eliminating, and disposing of hazardous materials. Process monitoring, along with operator instructions, must be documented and the documentation is to be accessible at work stations. All process changes that contrast with the control plan must be approved by the customer.

Inspection and Testing

Under QS 9000, the only acceptable levels for attribute sampling plans are zero defects. All process activities should incorporate defect prevention techniques such as statistical process control (SPC), error proofing, or visual controls. These should virtually eliminate the need for techniques that detect defects. AIAG offers a manual to help implement statistical techniques. Titled *Fundamental Statistical Control Reference Manual*, it contains a good deal of information on continuous improvement, SPC charts, and process capability.

If, despite all safeguards, a nonconformance appears, either internally or externally, certain problem-solving methods are called for in QS 9000. The procedures address effective handling of customer complaints and reports, investigation of nonconformity causes, determining appropriate corrective action, and applying controls to ensure that the corrective action is indeed effective. Returned parts must be analyzed and corrective action taken to prevent recurrence.

More information on QS 9000 can be obtained from the Automotive Industry Action Group, 26200 Lahser Road, Suite 200, Southfield, Michigan 48034.

HELP FROM ISO

ISO provides guidelines for use in implementing the ISO 9000 standards for a generic quality system. Additional guidelines are planned for release over the next several years. Many will be adopted by various countries as national guidelines, such as for the Q9000 standards in the U.S. The following are key guidelines that may be useful in carrying out the different inspection activities required by ISO 9000 or, in the U.S., Q9000 standards.

ISO 8402: "Quality Management and Quality Assurance - Vocabulary." ISO 8402 is intended to clarify and standardize the major terms used in discussing and documenting quality systems, which include inspection activities. It is directly aimed toward those quality systems based on the ISO 9000 standards. In the past, definitions for specific quality terms have come from various sources and did not always agree. This guideline, which provides definitions officially sanctioned by ISO, contains terms used in many situations—quality manuals, procedures, work instructions, or in everyday conversations. When these terms are used, they must be consistent with the guideline to avoid confusion.

ISO 10013: "Guidelines for Developing Quality Manuals." The purpose of this guideline is to provide help in developing a quality manual to support an ISO 9000-based quality system. Both the U.S. national Q9000 standards and the automotive industry QS 9000 refer to it for obtaining help. A quality manual is a document describing a company's quality policy and its quality system. Forming the foundation for procedures, quality plans, and detailed work instructions, the quality manual may be used by a company for:

- communicating the company's quality policy, procedures, and requirements;
- describing and implementing an effective quality system;
- providing improved control of practices and making assurance activities possible;
- providing documentation that can be audited;
- providing continuity of the quality system through changing circumstances;

- training personnel in the quality system requirements; and
- showing compliance with ISO 9000-based standards to external sources.

Sometimes, the quality manual may refer to just part of a company's activity. One example is when only some products are made under ISO 9000 requirements. Generally, the quality manual is relatively short and concise—25 to 30 pages. For help, ISO 10013 addresses such topics as preparing a quality manual and approval, issue, and control.

ISO 10011-1: "Guidelines for Auditing Quality Systems–Part 1: Auditing." ISO 10011-1 provides help, in general, for auditing quality systems based on ISO 9000 standards. Basic audit principles and practices are addressed. Help is also provided for establishing, planning, carrying out, and documenting audits. The guideline includes:

- definitions;
- audit objectives and responsibilities;
- audit initiation, planning, executing, and documenting;
- audit completion; and
- corrective action follow-up.

ISO 10011-2: "Guidelines for Auditing Quality Systems–Part 2: Qualification Criteria for Quality System Auditors." ISO 10011-2 aims to provide minimum criteria for those required to do quality audits. It also offers a method by which an auditor's compliance to this criteria can be evaluated and maintained. It includes:

- definitions;
- education;
- training;
- experience;
- personal attributes;
- management capabilities;
- maintenance of competence;
- language;
- selection of lead auditor;
- evaluation of candidates; and
- national auditor certification.

ISO-10011-3: "Guidelines for Auditing Quality Systems–Part 3: Management of Audit Programs." ISO-10011-3 provides help in managing the audit process. Contents of this guideline include definitions, managing an audit program, and a code of ethics. More information is included in the ISO guidelines available from the American National Standards Institute, 11 West 42nd St., 13th Floor, New York, NY 10036 or the American Society of Quality Control, 611 E. Wisconsin Ave., P.O. Box 305, Milwaukee, WI 53201-3005.

HELP FROM PROFESSIONAL SOCIETIES AND OTHER PUBLISHERS

There has been widespread interest in quality for the past few years, encouraged by the publication of many books and videotapes. These offerings can be useful in carrying out the quality system and, more specifically, inspection activities. Professional societies, like the Society of Manufacturing Engineers (SME), have catalogs listing their offerings. So do other publishing houses. Categories of books and videotapes at SME include:

- continuous improvement;
- design for manufacturability;
- manufacturing strategies;
- geometric dimensioning and tolerancing;
- production management; and
- Quality/TQM/ISO 9000.

Bibliography

Advanced Product Quality Planning and Control Plan. Southfield, MI: Automotive Industries Action Group (AIAG), 1994.

Fundamental Statistical Process Control Reference Manual. Southfield, MI: Automotive Industries Action Group (AIAG), 1992.

Guidance on the Application of ISO 9000/ASQC Q90 Series Quality System Standards. Department of Defense/NASA: MIL-HDBK-9000/NASA-HDBK-9000, 1994.

Guidelines for Auditing Quality Systems - Part 1: Auditing. International Organization for Standardization: ISO-10011-1:1990(E).

Guidelines for Auditing Quality Systems - Part 2: Qualification Criteria for Quality System Auditors. International Organization for Standardization: ISO-10011-2:1991(E).

Guidelines for Auditing Quality Systems - Part 3: Management of Audit Programs. International Organization for Standardization: ISO-10011-3:1991 (E).

Guidelines for Developing Quality Manuals. International Organization for Standardization: ISO-10013.

Hodson, William K., editor in chief, *Maynard's Industrial Engineering Handbook, 4th ed.* New York: McGraw Hill, 1992.

Juran, J.M., editor, *Quality Control Handbook, 4th ed.* New York: McGraw Hill, 1988.

Juran, J.M. and Gryna, Frank M. Jr., *Quality Planning and Analysis, 3rd ed.* New York: McGraw Hill, 1993.

Juran, J.M., *Juran on Leadership for Quality.* New York: The Free Press, 1989.

Malcolm Baldrige National Quality Award 1995 Award Criteria. Gaithersburg, MD: National Institute of Standards and Technology, 1994.

Peach, Robert W., editor, *The ISO 9000 Handbook, 2nd ed.* Fairfax, VA: CEEM Informational Services, 1994.

Pince, Bruce W., et al, *Automotive Quality Requirement Standardization.* Milwaukee, WI: ASQC Quality Congress Transactions-Nashville, 1992.

Potential Failure Mode and Effects Analysis Manual. Southfield, MI: Automotive Industries Action Group (AIAG), 1993.

Potential Failure Mode and Effects Analysis in Design (Design FMEA) and *Potential Failure Mode and Effects Analysis in Manufacturing and Assembly Processes (Process FMEA Reference Manual).* Warrendale, PA: Society of Automotive Engineers, SAE J1739, 1994.

Production Part Approval Process. Southfield, MI: Automotive Industries Action Group (AIAG), 1993.

Quality Management and Quality Assurance - Vocabulary. International Organization for Standardization (ISO/DIS) 8402, 1992.

Quality System Requirements. Southfield, MI: Automotive Industry Action Group (AIAG) QS-9000, 1994.

Salvendy, Gavriel, editor, *Handbook of Industrial Engineering.* New York: John Wiley, 1992.

Tooling & Equipment Supplier Quality Assurance Guideline. Southfield, MI: Automotive Industry Action Group (AIAG), 1988.

Inspection and Measurement

Winchell, William, *Continuous Quality Improvement: a Manufacturing Professional's Guide*. Dearborn, MI: Society of Manufacturing Engineers, 1991.

6

Inspection Software

In the past decade, the use of computers by industry has grown immensely, with much of this growth in personal computers (PCs). During this time, there was no comparable increase in the use of main frame computers.

Many PCs sold today are more powerful than the main frame computers in use only a few years ago. Yet, the price of PCs for an entire company may be a fraction of what a main frame would cost. Flexibility and wide availability of off-the-shelf software add to the appeal of PCs.

Inspection activities need to be computer based to meet the wide variety of demands placed on them. Today, applications can be found on both main frames and PCs. There is also a middle of the road called a minicomputer used particularly in medium-size companies. Many of these minicomputers are used to drive production control applications. Often, inspection activities are part of the production control applications.

Software to support inspection activities is available for any of the three major types of computers, as illustrated in Figure 6-1. That used on main frames or minicomputers is likely to be customized by internal data processing professionals or outside consultants. Software used on PCs is mainly used as it was purchased, without any significant modifications.

CHARACTERISTICS OF INSPECTION SOFTWARE

Software that supports inspection activities has common characteristics relating to the knowledge base. A knowledge base is composed of documents and data stored in the computer's memory. In nearly all applications, the knowledge base is constantly changing. To adapt to this, software should allow the entry of documents or data into the computer memory. This may be by hand, such as a keyboard, or by receiving signals generated by another device or computer system. The software should also allow editing and modifying of the documents or data. Calculations should be

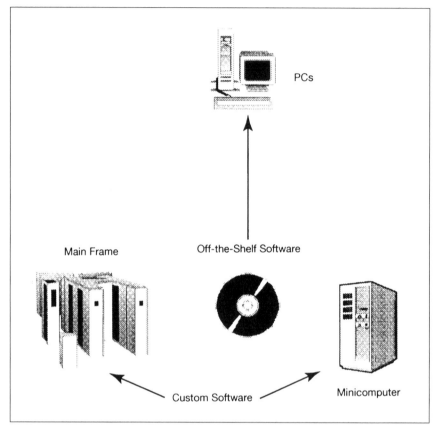

Figure 6-1. Major types of computer software used by various computers to support inspection.

done as necessary to turn the data into useful information. To make these calculations, information in the knowledge base may be combined with data that is input. The output information should be formatted to meet the needs of those using it. Sometimes, output may be displayed on a monitor. For other situations, a computer printout may be needed or the information may be sent to another computer system or device. The software should help protect the integrity of the knowledge base and the system.

Advantages of Inspection Software

Many demands are placed on inspection activities to meet customer requirements. Documents, such as procedures and work instructions, must be up-to-date. Easily retrievable records must be kept for the many facets of producing and checking products in production.

For most companies, doing these tasks manually is no longer feasible. There are just too many advantages to using software for inspection. The knowledge base can be kept current easily, and this up-to-date information can be readily accessed by anyone on the network. Checking and editing can be performed quickly and with greater accuracy. Calculation, summation, and preparation of output reports can be automated, eliminating what otherwise could be a laborious and lengthy effort. Accessibility can be limited only to those responsible for changing specific documents, thereby protecting the system's integrity.

By using software to support inspection activities, procedures can be designed that are coordinated with other computerized systems in a company to achieve high degrees of effectiveness. In this way, the procedures can be more directly responsive to the needs of the information users. But the key to using inspection software effectively is to be certain that the data is kept current.

Global Inspection Software

Software supporting inspection activities could be viewed broadly on a global basis as in Figure 6-2. The advantage is that the basic intent of the software supporting inspection modules will not likely change as time passes. However, the software within each module will change a great deal due to new offerings and updates. Also, it is expected that new offerings will grow broader in scope and handle the intent of more than one module. Software will probably change rapidly from standalone to integrated and networked versions with many other applications. The modules, from a broad or global perspective, can be viewed as providing support for inspection in various ways.

Planning

The planning module supports inspection by helping to plan what should be done in prototype and production processes. A major part of this is to generate and maintain the quality plan or control plan, as it is sometimes called. Software for getting the inputs, such as a process flow chart, DFMEA, and PFMEA, into the control plan may be in this group.

This module also may be integrated with the documentation module providing a means of developing procedures and work instructions. Integration with the problem solution module could provide valuable opinions on the production processes and customer experiences to prevent and correct problems.

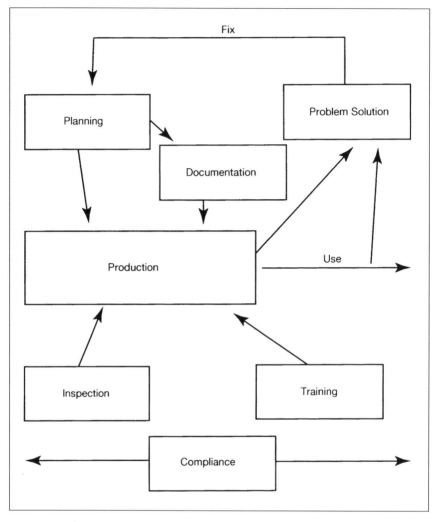

Figure 6-2. Global view of software that supports inspection activities.

Documentation

This module preserves information vital to running production processes. The quality policy, procedures, control plan, and work instructions are stored in its knowledge or data base to be prepared and upgraded by inspection software integrated with that in the planning group. Software allows this data base to be updated readily and controls can be used to assure the integrity of the stored information.

By integrating with the inspection group, data or records could be imported and stored in the documentation knowledge base. Part of this knowledge base also may contain such things as product designs, process designs, math data, or specifications. This information may be imported from an existing design data base or accessed through a network.

Inspection

Keeping track of the product is the major task of the inspection module. Often this is done by using an existing production control data base through a computer network. Inspection data, regularly entered in the data base, keeps the status of products up-to-date.

Other software in this module may concern such things as sampling, test preparation, process capability analysis, and statistical process control. This is also where the various strategies in the control plan are integrated in the process. Software outputs may be integrated with the documentation module and stored in the knowledge base for future reference. Certain outputs from transducers also may drive process control devices or automatic inspection in the production processes.

Training

Training is a strategic issue for any company. It is essential that each person be trained to do the job expected. Software in this module provides help in preparing a needs analysis and evaluating progress toward meeting goals. It also provides a means of periodically evaluating training effectiveness. How each individual in a company is progressing toward training objectives is also a key feature of this module.

Problem Solution

Corrective action and implementing prevention strategies are important to the long-range viability of a company. The module for problem solution takes data from production processes and customers as they use the products. The challenge is to apply software to convert data into useful information so that problems can be identified and solutions sought.

Software in this module includes such packages as problem solving, statistical methods, reliability, quality cost, surveys, and design of experiments. The module is tightly integrated with planning and production so changes can be quickly made. Valuable feedback is provided for correcting and preventing problems.

Compliance

This module, which provides a method for assuring that the quality system is operating properly, supplies software that can support the audit function. Software may include such things as auditing, ISO 9000, and supplier quality assurance. Results encourage changes useful throughout the quality system, including inspection activities.

Planning for Software

Determining what software supports an inspection activity is critical. It would be inadvisable to pay for a custom program when something off-the-shelf would do the job. On the other hand, it does not make any sense to get an off-the-shelf package and then find out it has some real short-comings. To avoid these mistakes, time must be taken to specifically determine what is needed.

Often complicating planning is the need to provide a system useful to many departments in a company. This situation may require:

- planning the system from an overall company viewpoint;
- entering documents and data from several departments that must be consolidated in a seamless knowledge base;
- a common knowledge base used by all departments in a company rather than standalone applications;
- updating the knowledge base in a timely fashion;
- accessibility of the programs and knowledge base at different locations in a company; and
- flexible output that meets the needs of all information users.

Certain questions should be asked during planning to find out what is needed. Some of these questions are:

- What is the intent of the system and the procedures?
- Who are the various users of the information and what are their needs?
- Who will enter documents and data and how will it be done?
- What are the report format requirements of the users?
- What controls will provide the necessary integrity?
- What is the existing system and what are the shortcomings?
- Can any part of the existing system be used?
- What software is available off-the-shelf and what are its short-comings?
- What are the hardware and networking requirements for the various alternatives?
- What is the timing for system start-up for the various options?
- What are the training requirements for the various choices?
- What are the cost and benefits of the options?
- Will the customer or third party registrars be satisfied by the alternatives?

Custom Software

Custom software to support inspection activities appeared soon after the availability of the main frame computer in the 1960s. Originally, programs were designed by the data processing department as part of production control efforts. Only large companies could economically justify this effort. Later, main frame manufacturers and third parties offered software packages usually modified for each application.

The intent of early programs was to track good and bad products through the manufacturing process. Good products were shipped to customers; bad were either reworked or scrapped. Data was entered on a computer form and keypunched on cards for processing on a card reader. The programs did calculations and printed summary sheets, driving such actions as adjusting production schedules. Much time was spent finding and correcting mistakes in the data, which was manually entered, to get the results right.

Data entry by terminals networked to the main frame started in the 1970s. This vast improvement over previous means of data entry provided timely and more accurate results. For some companies, this progressed to electronically entered data from sensors on the equipment.

The wide availability of minicomputers during the 1980s made production control computer systems feasible for smaller companies. Also, larger companies could now have a dedicated computer that had no other priorities. Modules to support inspection were provided by the computer manufacturers. Later, many third parties offered a wide selection of software. Currently, these programs are either customized internally or by outside consultants to make them fit better with the needs of each specific company.

Off-the-shelf Software

Over the years, inspection reports were added to make the main frame and minicomputer systems more useful. However, their usually fixed formats made critics doubtful about their utility for solving problems. Often these reports were left unread because they did not provide information in an easily accessible format. Storage and easy retrieval of documentation, such as procedures and work instructions, also became a major issue in the late 1980s.

Nevertheless, many users of inspection information sought more flexibility in programs and reports for inspection. Also, increased capability to do statistical studies was needed. These needs, and others, led the migration to PCs in the 1980s. Widespread availability of networks for linking PCs came about in the early 1990s, allowing the same information and capability to be available at multiple sites. Although custom software is sometimes used for PCs, much of the software is purchased off-the-shelf and used without modification.

For more than 10 years, the American Society for Quality Control (ASQC) has published an annual directory of software concerned with quality. Each year, the available software has grown in number and in choices. The 1995 directory had 526 software packages from 245 companies.

These programs can be run on many computers. Although PC applications dominate the offerings, other operating systems are listed for some programs, including DEC ALPHA, DEX VAX/VMS, UNIX, IBM main frames, Macintosh, and others. For the most part, the software is off-the-shelf and can be used without customization.

Software is constantly changing and is very dynamic in nature. Some changes are reflected in upgrades; others are incorporated in new offerings that reflect new ideas or concepts. Currently, ASQC groups software concerning quality into 26 categories. Some software naturally falls into several categories as it has multiple purposes. Those categories that may have software to help support inspection activities are:

- auditing;
- Baldrige Award;
- benchmarking;
- capability studies;
- data acquisition;
- design of experiments;
- inspection;
- ISO 9000;
- management;
- problem solving;
- quality costs;
- quality function deployment;
- reliability;
- sampling;
- statistical methods;
- statistical process control;
- supplier quality assurance;
- survey techniques;
- Taguchi techniques;
- test preparation; and
- training.

Bibliography

"1995 Software Directory." Milwaukee, WI: *Quality Progress,* March 1995.

Banks, Jerry, *Principles of Quality Control.* New York: John Wiley, 1989.

Brumm, Eugenia K., "Managing Records for ISO 9000 Compliance." Milwaukee, WI: *Quality Progress*, January 1995.

Hodson, William K., editor-in-chief, *Maynard's Industrial Engineering Handbook, 4th ed*. New York: McGraw Hill, 1992.

Juran, J.M., editor, *Quality Control Handbook, 4th ed.* New York: McGraw Hill, 1988.

Juran, J.M. and Gryna, Frank M., Jr., *Quality Planning and Analysis, 3rd ed.* New York: McGraw Hill, 1993.

Salvendy, Gavriel, editor, *Handbook of Industrial Engineering.* New York: John Wiley, 1992.

Weber, Ernest G., "Computers, Metrology and Manufacturing." Carol Stream, IL: *Quality*, January 1992.

Winchell, William, *Realistic Cost Estimating for Manufacturing, 2nd ed*. Dearborn, MI: Society of Manufacturing Engineers, 1989.

Part 2

MEASUREMENT

General Measurement Concepts

This chapter begins the second part of the book, and concerns measurement, specifically dimensional measurement, an area of high interest for the manufacturing professional. Companies engaged in manufacturing use measurements to find if products are within specifications. Measurement provides facts for assuring quality, as well as for general improvement efforts.

Measuring length also has a special role in measurement technology because nearly all analog measurements can be reduced to measurements of length. For example, such things as angles and areas can be calculated by using length measurements. Also, deformation due to a force is measured as a change in length. Instruments such as pressure gages also rely on measuring a change in length for their readings.

MEASURING METHODS

One of the earliest ways to measure length reliably is still being used today. It is the ruler. Depending on the type of ruler, an accuracy of 0.04 inches (1 mm) can be expected. Vernier calipers may have an accuracy of 0.002 inches (0.05 mm), while micrometers are viewed as better with an accuracy of 0.0004 inches (0.01 mm). In 1915, the American War Department established Johansson gage blocks as the standard for workshops manufacturing war material. Johansson's system was patented in Sweden and had what was called progressive tolerances that were a small fraction of a micrometer. All would agree that Johansson's system made possible the steady progress to more accurate measurements.

Interferometer

Today, the most accurate way known to measure length at many locations is with an interferometer. The standard meter, which forms the foundation of measurement, was once based on a given wavelength of light. In 1983, the definition of the standard meter was changed to the length of the path traveled by light in a vacuum during 1/299,792,458 of a second.

Current Developments

The science of dimensional measurement is still being developed. Emphasis on improving measurement has produced the rapid deployment of in-process gaging, feedback for process control devices, and smaller tolerances to meet customer needs. For tool and gage makers who traditionally use 10% of the product tolerances in making their applications, the need to get better measurement capability is even more pressing.

The challenge lies in measuring what is happening in the real world. There is nothing perfectly flat, round, or square. Supposedly, duplicate parts actually vary in size, location of features, and form. Adding to the challenge is error inherent in any measurement method. Making the problem even more severe is the need for even more accurate parts. When James Watt was boring the cylinders in his steam engine, he proudly remarked that he attained uniformity of the diameter within a "worn" shilling. It would be difficult to find applications in manufacturing in which this little precision would be acceptable today.

MEASUREMENT PROCESS

One could view measurement as a process in which the thing to be measured is compared with a predetermined reference or standard. The measurement involves both a number and a unit of measure—for example, 1 inch (25 mm). More formally, measurement is defined by ISO as a set of operations having the object of determining the value of a quantity, like a dimension. The value will not always be the same each time a measurement is made.

Like other processes, measurement can be seen as a set of interrelated sources, activities, and influences that produce a measurement. Influences such as the environment may add variability to the process.

The measurement process has inputs, outputs, and steps. In simplistic terms, inputs are the articles to be measured, the operator, procedures, and a measuring instrument. Inputs also include a reference standard for calibrating the instrument and a proper environment for doing the measurement. The output is, of course, the measurement.

At one extreme, the process may consist of various operators using informal procedures and general instruments in the plant. In a more formal setting, it may encompass trained calibration laboratory technicians using a measuring system that follows a detailed procedure in a controlled environment.

In a plant, a measurement could be used as a comparison to the specifications of the article being measured. The basic steps in the process may include:

- define the quality characteristic to be measured;
- determine the units for the measurement;
- establish what environment is necessary;
- choose the method to be used for the measurement;
- select the appropriate instrument to be used;
- establish the proper setup for the article to be measured;
- document the procedure to be used;
- calibrate the instrument to a reference standard if required;
- do the measurement according to the procedure; and
- report the measurement.

Measurement Errors

Since variation is inherent in the measurement process, the measured value is not equal to the actual dimension of a part. Such errors, when large, may have dreadful consequences for a company. In reality, a company may not know whether a product is okay or not. In addition, controls for the process may not be based on what is really happening. Bad parts could be shipped to customers, causing much ill will. On the other hand, good parts may be wrongly classified by a company and thrown out as scrap.

PRECISION AND ACCURACY

To grasp the meaning of measurement error, an understanding of the nature of precision and accuracy is needed. These terms, unfortunately, are commonly misunderstood and confused with each other. Adding to this confusion is that the terms are used in several ways by those concerned with measurement.

Precision

The degree of agreement among repeated individual measurements of the same sample over a long time can be viewed as precision as shown in Figure 7-1. Of course, these individual measurements must be repeated under the same conditions in a statistical study. Normally, the standard deviation of the repeated individual measurements is used to quantify the precision or spread of the measurement process.

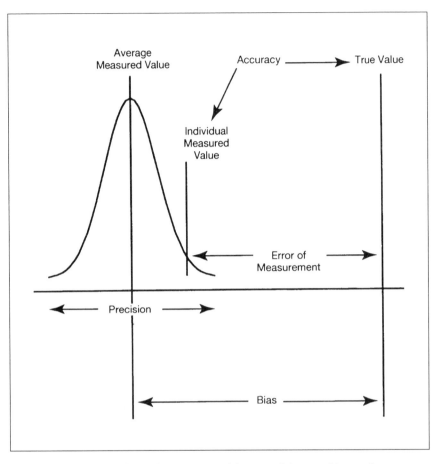

Figure 7-1. Diagram illustrating accuracy, bias, precision, and true value.

Uncertainty of Measurement

One term closely associated with precision is uncertainty of measurement. There is wide agreement that the uncertainty of a measurement should be reported in two parts—that which can be determined statistically and that which cannot. The first part is equivalent to precision and is found and stated in a like fashion.

Uncertainty of measurement is defined as the range in which the true value of a measurement is estimated to lie, generally within a given confidence level. A confidence level may be expressed as a standard deviation, like precision, or the half-width of an interval having a stated level of confidence.

The second part of uncertainty is expressed in the same way, but can only be determined through the judgment of those concerned. The judgment is largely based upon experience, and cannot be found through a statistical study as with precision. An example would be gage blocks calibrated by a laboratory. Although, close control to 68° F (20° C) is sought, the temperature will vary minutely. The exact temperature at the time the gage block in question was calibrated is unknown. But the possible variation in temperature during this time period is known. Based on this, the possible variation in the dimension of the gage block could be calculated. What must be known are the probable range of temperature when the calibration was done and the coefficient of thermal expansion of the gage block.

Collectively, the two parts comprising uncertainty represent a worst-case scenario. One could view uncertainty, in this manner, as the maximum expected variation of a measuring system considering known factors. But, if an unknown factor becomes influential, this worst-case estimate of variation may be exceeded. Often it is an unknown factor that causes measurement capability to deteriorate.

Repeatability

This term is commonly used to express the precision or variability due to the measuring equipment. Although only part of the measurement variation, it could be a major contributor. To find repeatability, a controlled experiment is usually done. Repeated measurements are made under the same conditions, which include the same operator, test procedure, and equipment. Studies are taken over a short period. Repeatability is used in judging the adequacy of the equipment component of a measurement system.

Reproducibility

The precision or variability due to the operator is commonly called reproducibility—often a major portion of the entire measurement variation. To determine normal reproducibility requires a controlled experiment that uses different operators, but keeps the same test procedure and equipment. Studies are taken over a short period. Reproducibility is used in judging the expected adequacy of the human variation in a measurement system.

Two Viewpoints Regarding Precision

Today there are two major, but seemingly divergent, viewpoints concerning precision. At first impression, it may appear that only one is correct. Those most vocal on either side of the discussion may try to lead others to this conclusion. Nevertheless, in fact, both viewpoints are equally important and critical to the future progress of manufacturing. The two viewpoints are called "Postulate of Imprecision" and "Deterministic."

Postulate of Imprecision

Proponents of this viewpoint trace its origin to Plato. Using principles of geometry, Plato proved that a circle would always have variability despite how precisely it was drawn. This notion of an ideal form and the difficulties associated with achieving it in practice have been discussed by both philosophers and engineers over the centuries.

Early in the 19th century, armament makers took Plato's hypothetical model into consideration when designing jigs and fixtures. For critical dimensions, gages also were designed keeping Plato's model in mind. Parts produced approximated those predicted by the model, allowing interchangeability on a large scale. This success influenced other companies to adopt a similar approach. World War II accelerated the notion of variability or imprecision in making parts. Ordnance manuals evolved into military standards for dimensioning and tolerancing of the variation inherent in manufacturing. Later, this movement inspired national standards and then international standards.

Today, most people concerned with manufacturing believe in the Postulate of Imprecision. They are convinced that variation in product geometry is inevitable and a strategy must be provided to control it. This strategy includes obtaining better methods of measurement to drive process control efforts and judge whether a product is in tolerance.

Deterministic Approach to Metrology

The deterministic school of thought traces its origin to the late John Loxham who closely studied automatic manufacturing processes during this century. The Loxham Principle states that an automated manufacturing process always operates perfectly; if it does not, it is not set up correctly.

Advocates of the Loxham Principle declare that exceptions to the deterministic laws of nature occur only at the subatomic level. Automatic manufacturing processes are said to operate only under deterministic natural laws.

A spokesperson on the Loxham Principle stated recently that all nonrepeatable variation observed in his experiences and those of others were systematic in nature and could be corrected. All one needs to achieve perfection, he argued, are time, money, and the right attitude.

Those favoring the Loxham Principle think that causes of systematic variation are limited enough to understand and correct. The causes are believed to include human actions, thermal effects, friction, lubrication, dirt, and variations in the supply of utilities.

Because of work on understanding precision, such as advocated by proponents of the Loxham Principle, much progress has been made. Perfection has not been reached yet, but it is always being sought. The process of seeking to be better is the backbone of continuous improvement.

There is an increasing number of applications that are closer to perfection. Examples are:

- drift between the tool and workpiece in a diamond turning machine is consistently less than 25 nm;
- the diameter in a microscope is within 50 nm when using a diamond boring operation; and
- magnetic recording surfaces are within 10 nm using float polishing technology.

These examples, while impressive, still require more precise manufacturing processes and measurement strategies. To support this need for precision, better manufacturing processes and measuring strategies must be adopted. This will take both an understanding of variation and the constant search for perfection. Both viewpoints will drive technology on the road to the needed continuous improvement.

Accuracy

One could broadly view accuracy as the closeness of a measurement to the true value of a characteristic. In practice, accuracy can be only closely approximated. This is because the true value of a characteristic is never known, even for reference standards. There are minute errors in the manufacturing of reference standards that are not known by those using them for calibration. For the manufacturing professional, the mystery of never knowing the true value of a characteristic is not a concern.

There are two ways commonly used to describe accuracy in more detail. The first way, illustrated in Figure 7-1, assumes that accuracy is based on a single measurement. Accuracy is defined, in a qualitative manner, as the agreement of an individual measurement with the true value of the characteristic being measured.

Error of Measurement

To quantify accuracy, the error of measurement is used. It is the difference between the individual measurement and the true value and is the actual divergence. Bias also is used in this process, and is the predicted difference, on the average, between the measurement and the true value. Bias, sometimes called systematic error, is the extent that an instrument is out of calibration. A systematic, or random, error remains constant or varies predictably and is an unwanted offset in the measurement process. Bias is the correction required to compensate for the systematic error.

Accuracy, addressed in this manner, is very complex because it is dependent on both the precision and bias of the measurement process. To properly define accuracy in this way, it is necessary to say it is precision, the bias or systematic error, and the form of the distribution of individual measurements about the average of the measurements.

The second way to describe accuracy assumes that it is based on repeated measurement, under the same conditions, of the same characteristic (Figure 7-2). The wide availability of CMMs today makes this way of

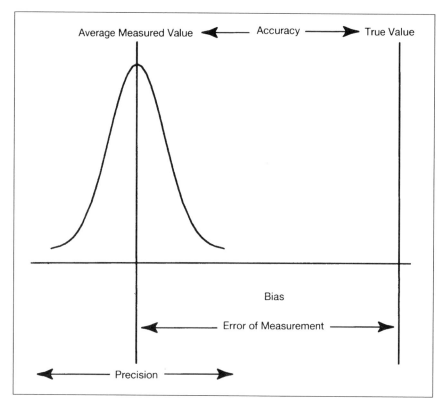

Figure 7-2. Statistical approach diagram showing accuracy, bias, precision, and true value.

finding accuracy viable using a statistical study. Accuracy is defined, in a qualitative manner, as the agreement of an average of multiple measurements with the true value of the characteristic being measured.

Accuracy, defined in this way, is simpler because it depends only on the bias of the measurement process. Bias, which is identical to the error of measurement, is the correction required to compensate for any systematic error when doing calibration.

COST OF MEASUREMENT

Striving for perfection could reach a point of diminishing economic returns for some applications. In making progress regarding precision, the tradeoff between the profit that will likely occur and the cost to gain the precision will continue to be a key issue in the future of manufacturing.

The costs to get better measurement capability will vary, depending on the application. Sometimes costs will fall when technology advances take place. At other times, such advances may be accompanied by an exponential rise in costs.

In retrospect, history builds a powerful case for constantly seeking greater precision in both technology and measurement. The latter half of the 20th century has had an explosive growth in products with greater precision. In only 50 years, the typical scale for measurement in high technology applications changed from 1/10,000 to 1/1,000,000 of an inch precision. Examples are plentiful, ranging from integrated circuits to jet plane turbine engines. Seeking greater precision in its product will undoubtedly help ensure a company's survival in the marketplace. The accompanying cost to get better measurement capability may be unavoidable for companies to stay in business.

Those seeking to estimate the cost of a measurement application should gather information on both the initial and the operating cost. Considering only initial cost when judging alternative methods for measuring sometimes may lead to the wrong conclusion. Identifying the operating cost up front prevents surprises when the application is finally running. Operating cost includes the expense of running the measuring equipment, and recording and reporting information.

MAJOR FACTORS AFFECTING MEASUREMENT

Factors influencing how well measurements are taken may be divided into five categories. They are: the reference standard, workpiece, instrument, operator, and environment.

The workpiece, for example, may affect the measurement through its surface finish, elastic properties, cleanliness, and thermal equalization. The instrument must be checked for its amplification; electric, optical, or pneumatic input; or if it has been deformed by heavy items. Of course, operators' variables include such things as skills and training. The factor that perhaps presents the most variables is the environment. Here you must consider such items as radiation from lights, heating components, drafts of air, clean surroundings, vibrations, etc.

Many more factors influence most measurement situations, and an attempt should be made to identify all of them. Unfortunately, undiscovered factors are often found later, after a problem in measurement is detected. A known factor can be evaluated and a sense of its importance established. Some factors are critical in a particular situation; others are unimportant and may be disregarded.

MEASUREMENT STRATEGY

Although measurements were made early in ancient history, the basis for a measurement strategy can be traced only to the 19th century. "Strategy" refers to a method to reach an objective. That objective is usually to decide if a product meets requirements.

Two Technologies

There are two separate technologies on which to base the strategy for measuring.

In the mid-19th century, mass production of vernier calipers led to the strategy *called plus/minus tolerances coupled with two-point measurements* (Figure 7-3). Many companies still use this today for quick checks. However, people using this method often assume wrongly that the part is square and flat. Measurements by this method may not be precise because different pressures may be used to secure the instrument to the workpiece. Ratcheting devices have been adopted to prevent this.

To produce muskets that were interchangeable, armament makers early in the 19th century used a strategy called *geometric tolerances and functional gages*. It was based on concepts that kept the Plato model in mind. For some characteristics, maximum and minimum material conditions were established that could be checked by gages, which were used by many companies to decide if a part was acceptable. Go/no-go gages were widely used during much of the 20th century, but are now being replaced by variable gages. In the mid-20th century, this approach was formalized in the evolving ANSI Y14.5 series of standards on geometric dimensioning and

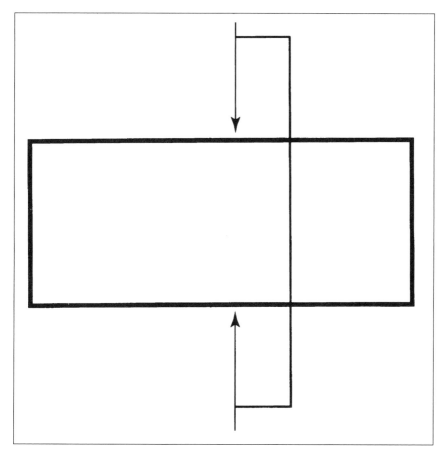

Figure 7-3. Early approaches to measuring featured two-point measurement.

tolerancing (GD&T). Although the earlier standards were said to apply to surface plate measurements, some argued that they really reflected gage practice. However, they were clearly not gaging standards in that there was not an associated measurement standard.

Did They Work?

Up to 1980, the two measurement technologies seem to have worked satisfactorily. Some credit the many skilled artisans during this period who solved the ambiguities in measurement with common sense reasoning.

But did the two approaches work well? Critics, through 20/20 hindsight, say that this was an illusion. The meaning of quality before this time did not focus on the customer, but rather on the factory floor.

For many companies, a product was considered good quality if it passed final inspection. Inspectors often practiced much leeway in making this decision. For example, when tolerances were not met, discretion was widely practiced. When the inspector did not want to decide, written deviations that required management signatures were used. There were many deviations in most companies.

Liberties also were taken with measurements. Those taking two-point measurements commonly assumed that the workpiece was perfectly flat and square. More pressure on a "mic" could get a different reading that may be more favorable. By oiling a gage, a tight workpiece could probably fit. The list of ways to change a measurement to something more favorable went on and on.

The foregoing scenario was widespread. Consumers tolerated the marginal product quality because they really had no choice. This changed when better quality products, often at a lower price, became available. Quality then became the dominant issue.

To meet customer needs better, technology aimed at gaining more precision was widely adopted during the 1980s. This included more reliance on computers for design and process control, and greater use of coordinate measuring machines (CMMs). Variable gaging also was adopted where attribute go/no-go gages previously did the job. Going along with this were tighter tolerances to give customers what they expected.

CMMs introduced a de facto third measurement technology in the 1980s that was widely adopted to get better quality. This is shown in Figure 7-4. The ability to create products with greater accuracy and precision improved during this period. However, three measurement technologies were now being used simultaneously in most companies. Inspectors soon found that the results from these methods did not concur because of ambiguities in the definitions of ANSI Y14.5. In addition, results from CMMs made by different manufacturers did not concur because there was not a standard mathematical algorithm for making calculations. To resolve this, it was agreed by the professional community that ANSI Y14.5 needed to be mathematized and generalized, and that measurement processes to assess conformance needed to be facilitated by ANSI Y14.5.

This agreement produced a new standard, ANSI Y14.5.1, dealing with the mathematical definition of dimensioning and tolerancing. The new Y14.5.1 was accompanied in a 1994 update of Y14.5. In 1996, a new standard, B89.3.2, will deal with dimensional measurement methods.

Some view these standards as a new era. Before this, dimensional tolerancing and metrology evolved as practices without strong theoretical underpinnings. The evolution will now continue upon a stronger base.

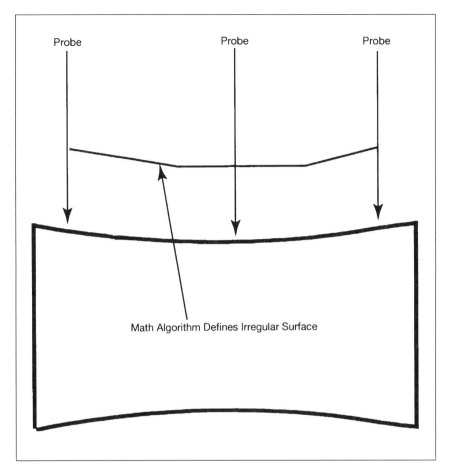

Figure 7-4. Latest measurement approach recognizes CMM technology, and that probes can find points for math algorithm that defines an irregular surface.

Bibliography

1995 Grolier Multimedia Encyclopedia. Windows/MPC Version 7.01, 1995.

Busch, Ted, *Fundamentals of Dimensional Measurement.* Albany, NY: Delmar, 1989.

Farago, Francis T. and Curtis, Mark A., *Handbook of Dimensional Measurement,* 3rd ed. New York: Industrial Press, 1994.

Grove, John W., editor, *Handbook of Industrial Metrology.* Englewood Cliffs, NJ: Prentice Hall, 1967.

Inspection and Measurement

Juran, J.M., editor, *Quality Control Handbook,* 4th ed. New York: McGraw Hill, 1988.

Juran, J.M. and Gryna, Frank M. Jr., *Quality Planning and Analysis,* 3rd ed. New York: McGraw Hill, 1993.

Salvendy, Gavriel, editor, *Handbook of Industrial Engineering.* New York: John Wiley, 1992.

Sprow, Eugene, "Challenges to CMM Accuracy." *Tooling and Production,* November 1990.

Srinivasan, Vijay, "Critical Dimensions." *Manufacturing Review*, V7 N4, December 1994.

Srinivasan, Vijay and O'Connor, Michael A., "On Interpreting Statistical Tolerancing." *Manufacturing Review*, V7 N4, December 1994.

Suresh, K. and Voelker, H. B., "New Challenges in Dimensional Metrology: A Case Study Based on Size." *Manufacturing Review*, V7 N4, December 1994.

Winchell, William, *Continuous Quality Improvement: A Manufacturing Professional's Guide,* Dearborn, MI: Society of Manufacturing Engineers, 1991.

Measurement Basics

Every measurement has three elements that must work in concert to ensure accuracy. These elements are:
- characteristic on the article to be measured;
- standard to which the measurement is to be compared; and
- an instrument to make the comparison.

Some measuring instruments, like the vernier caliper, have a built-in measurement standard. This makes it simpler to compare the characteristic on the piece being measured to the standard of measurement, which is the scales on the instrument.

All measurements, no matter the size, must start at a reference point and end at a measuring point. Many measuring instruments, again like the vernier caliper, use one jaw as the reference point. The other jaw, when properly tensioned on the article to be measured, is the measuring point.

ARTICLE TO BE MEASURED

An article having a characteristic to be measured must have a reference point from which the measurement starts. This is called a datum and is usually a geometric plane noted on the drawing. A datum establishes the reference frame on a part to be used in manufacturing, measurement, and assembly. Datums were illustrated in Figure 3-1.

The orientation, or direction, and size of the characteristic's dimension should be on the drawing, ending at a measuring point. The measuring point must be distinguishable and may be a geometric plane. Geometric dimensioning and tolerancing (GD&T) is increasingly used for dimensions. With GD&T, consistent interpretations are more likely among the parties involved.

Ideally, the article for which a characteristic is to be measured:
- is rigid and not subject to deformation that will affect the measurement;

- is produced using the same datums as specified on the drawing; and
- has datums that are structurally sound and will not deform significantly.

Articles that are not rigid, such as parts made from rubber, are typically supported in a fixture or gage when measured. Other means may be used, such as freezing the part.

MEASUREMENT STANDARDS

All measuring systems, even those existing today, were developed through subjective decisions. Early measurements were often established by a monarch, based on a part of his body. As civilization progressed, the systems were developed with more discussion and wider consensus among those concerned. Nevertheless, they were still subjective and had no natural basis. On reflection, one could say that each system may have been reasonable for the time in which it was used.

Evolution of Measuring Systems

The earliest recorded system of measurement was based on the Egyptian cubit, which quickly spread through the ancient world. Without such a system, the pyramids could not have been built. Its wide acceptance by other nations at the time also helped promote trade.

The metric system, completed in 1799 by a group of French scientists, is the most important system of weights and measures today. The U.S. Congress, in the Act of July 28, 1866, established the metric system as the lawful measure for any contract or dealing or pleading in any court throughout the U.S. This Act also tied the English system of measurement, based on the yard and pound, to the metric system by providing ratios for conversion. Interestingly, only the metric system was ever officially recognized as the national standard of the U.S. However, although it was legalized, its use was not made mandatory.

In 1901, Congress founded the National Bureau of Standards (NBS), which was charged with developing and maintaining national standards and carrying on a broad program of research in the measurement of physical quantities.

NBS worked with international bodies to further this effort on a worldwide basis. Such a body is the International Bureau of Weights and Measures (BIPM) in Sevres, a suburb of Paris. BIPM houses the international prototype meter used as an international standard for many years. In this way, the national standards were based on international agreements.

No regulatory powers were given to NBS; it was expected to maintain a standards and calibration service for industry and others that needed it. NBS, recently renamed the National Institute of Science and Technology (NIST), calibrates master measurement standards for nominal fees. Industrial laboratories are encouraged by NIST to calibrate their own working standards.

Although not legally the national standard, the English system of measurement was the de facto standard in the U.S. Various unsuccessful attempts were made during the 20th century to convert to the metric system. For example, an underwhelming attempt was started in the 1970s to display traffic signs in metric terms.

The English system of measurement flourished in the U.S., but its use in manufacturing has decreased dramatically during this past decade. The major reason is that more U.S. companies are engaged in multinational competition, where the metric system is routinely used. Critics of the English system feel that industry will lead the rapid conversion to metric in the U.S.

Requirements of a Measuring System

Measuring units must be consistently understood by all parties concerned. The metric system provides this advantage. All other systems, such as the English measuring system, are defined relative to the metric system.

Measuring units must be convenient and of practical size. One attractive feature of the metric system is that larger or smaller quantities are expressed as powers of 10. For example, a kilometer is 1,000 meters. Likewise, a centimeter is 1/100 of a meter. This must be contrasted with English measurement system units, such as the foot, which has 12 inches.

Most would agree that the metric system is more convenient for expressing and performing calculations. There would also likely be agreement that dealing with a single unit and its multiples provides a great deal of practicality in assessing the relative size of a quantity.

The SI System

In 1960, the International System of Units (SI) metric system was designated by the General Council on Weights and Measures (CGPM) as the international system of units. Over 45 member countries, including the U.S., send delegates to the CGPM, which has met periodically since 1889.

CGPM also has oversight of the BIPM. International standards have been maintained by BIPM since the late 1800s and are the references for the national standards of many countries. Costs for BIPM are shared by the member countries.

Now, the kilogram is the only SI unit kept at BIPM still defined as an artifact. The six other units have definitions that do not rely on an artifact. For example, the meter no longer depends on the distance between two marks on the international prototype meter. It is now defined as the speed of light. Since 1983, it has been the length of the path traveled by light in a vacuum during an interval of 1/299,792,458 of a second.

Originally, SI had six standard, or fundamental, base units. These concerned length, mass, time, temperature, matter, and current. CGPM added a seventh standard or fundamental base unit in 1971 for luminous intensity. Figure 8-1 illustrates the SI fundamental units that are the meter, kilogram, second, ampere, kelvin, mole, and candela.

Physical Quantity	Standard Unit	Symbol
Length	Meter	m
Mass	Kilogram	kg
Time	Second	s
Temperature	Degrees kelvin	K
Current	Ampere	a
Luminous intensity	Candela	cd
Matter	Mole	mol

Figure 8-1. Fundamental units of the SI system.

Only seven SI units are needed to measure all physical quantities completely with the techniques known today. It is a complete set of units since all other necessary units of measure are derived as a combination of the fundamental SI units. In a real sense, the SI units are also irreducible since no SI unit can be derived from any combination of the other SI units. Another SI feature is that larger or smaller quantities are expressed as powers of 10 for all the units. Figure 8-2 illustrates this relationship and the prefixes used.

Challenge to the Manufacturing Professional

For the manufacturing professional, the conversion to metric from the currently used system of units poses both a challenge and a source of confusion. For example, much equipment in the U.S. was made using the

Multiple	Prefix	Symbol
1 000 000 000 000 000 000	exa	E
1 000 000 000 000 000	peta	P
1 000 000 000 000	tera	T
1 000 000 000	giga	G
1 000 000	mega	M
1 000	kilo	k
100	hecto	h
10	deka	da
0.1	deci	d
0.01	centi	c
0.001	milli	m
0.000 001	micro	μ
0.000 000 001	nano	n
0.000 000 000 001	pico	p
0.000 000 000 000 001	fernto	f
0.000 000 000 000 000 001	atto	a

Figure 8-2. Prefixes used in the SI system to show magnitude.

English system of units. It will be a long time before it is replaced with equipment made under the metric convention. Also, many present product designs use the English system of units. Meanwhile, countless conversions and roundoffs are necessary.

In some countries systems are not purely metric, but somewhat hybrid in nature. They evolved from several measurement systems that had different units. For example, tires in the past have been sized in inches no matter where they were made in the world. Another example is that the inch is widely used in Germany and is called a "zoll." The list goes on and on. Because of this, and other subtle reasons, there are many ways that metric systems are used that in practice do not exactly match. All are very close to each other, but have slight departures so they can, for example, fit sizes that do not have perfectly aligned conversions.

This situation in the U.S., and other countries, forces what is often called "soft metrication," meaning that a metric quantity may be stated for a dimension with an actual size that slightly differs. Take, for example, a product that is already designed in English units. To be sold internationally, this design must be converted to metric units.

For example, a clearance in a part may be 0.040-inch (1.016-mm) minimum. There is a natural tendency by some to round off any metric number to a whole value. Rounding off to 1 mm minimum, which is 0.03937 inch, may be 0.0006 mm too narrow. Depending on how critical the clearance is, specifying the dimension as 1.017 mm minimum may be better, although it may look odd.

Dealing with this will probably require much thought and research before making adjustments and needed compromises in product and process designs. Coordination will be required to assure that a part made in the U.S. works, for example, with a part made elsewhere. The conversions from existing systems in each country must be compatible for both parts to work in unison.

MEASUREMENT INSTRUMENTS

Normally, equipment for measuring consists of instruments, any auxiliary apparatus, standards for comparison, reference materials, and the procedure or work instructions to do the measurement. Humans involved with making the equipment work may be included as part of this structure.

Certain characteristics are important in selecting an instrument for a measurement application. Requirements for each characteristic must be carefully considered for each application.

Accuracy. An instrument may have an accuracy specified as 1%. This is the error that may be experienced when using this instrument. The error is the difference between the true value of the variable being measured and what is shown by the instrument. Although there are many ways of expressing accuracy, most of the time it is given as a percent of the full-scale reading. Gage accuracy that is too large may be due to such things as the gage not being properly calibrated.

Linearity. If there are no differences in accuracy values through the specified measuring range, the measuring instrument is linear. If it is not linear, the relationship is represented by a curve and not a straight line. To have proper measurements, a linear relationship must normally exist throughout the range that measurements will be taken. In certain situations, a measuring instrument can be calibrated to account for relationships that are not linear.

Precision. Repeatability is the measure of an instrument's precision. It is the agreement of a series of repeated measurements, all made under the same conditions.

Sensitivity. The ability of a measuring device to detect small changes in a quantity being measured is called sensitivity. It is largely used in describing analog measurement signals. For example, take an instrument with a moving pointer. Sensitivity is how much a measurement must change before there is a detectable change in the position of the pointer. Sometimes this characteristic is called the threshold of response for the instrument.

Amplification. The ratio of the output of the measuring instrument to the input from a probe or sensor could be viewed as amplification. Amplification that is too low conceals variation in measurements. Measurements could be equally misleading if amplification is too high.

Resolution. When an instrument has a digital readout, the term resolution is more appropriate than sensitivity. It is the minimum change in a measurement signal that will change the value of a digital readout.

Concerning an analog needle readout, some view resolution as the ratio of the width of one scale division to the width of the needle. The needle width must be less than one scale division to make a reading that clearly separates the divisions on the scale.

Hysteresis. Hysteresis can be described as a lagging effect. When an operation is completed, for example, the return is not quite at the starting point. Unless compensated for, the next operation will not start in the right place. For a measuring instrument, this may be reflected in a measurement being different when going up scale as opposed to down scale. Or, it could be that the next measurements differ in progressively larger amounts.

Discrimination. Discrimination is the fineness of a scale on an instrument. On a micrometer, the vernier scale has a discrimination of 0.0001 mm. However, getting measurements having an accuracy of 0.0001 mm with a micrometer is doubtful.

Readability. Readability is a qualitative measure of the ability to convert the output of a measuring device to a meaningful number. In other words, can the output be read adequately? A digital readout, for example, improves the readability of a micrometer having a vernier scale. In a computer application, the output of a measuring device must be read by a software program.

Parallax. The angle at which someone views, for example, a gage having a needle could introduce parallax error. Depending on the angle of view, the reading could be low or high.

Specified Measuring Range. The specified measuring range is based on the limits of measured values that the instrument is capable of reading. It is important that the dimension of what is being measured fits inside the specified measuring range for the instrument.

Calibration. Calibration is finding the relationship between measured values and a reference standard for an instrument.

Adjustment. An operation done on a measuring instrument so that it performs correctly and without bias is called an adjustment. This is normally done based on calibration results. There are also characteristics that are important from a long-range viewpoint. They are as important as those from the immediate viewpoint. Requirements for these also should be carefully established, considering the application.

One is stability, the ability of a measuring instrument to keep a characteristic classified as immediate, constant over time. Many view stability as the dependability of a measurement process. It is important to recognize that stability is the ability of the measuring system to resist deterioration. Another requirement is drift, the slow change over time of a characteristic classified as immediate. It causes a measuring instrument to be unstable.

Availability. The percent of uptime for a measuring application is availability. Specifically, it can be calculated by:

$$\% \text{ Availability} \quad \frac{\text{Hours Uptime}}{\text{Hours Downtime} + \text{Hours Uptime}} \times 100$$

Selecting measuring instruments with a larger percentage of availability assures less disruption in measurement efforts.

Maintainability. This is an indicator of how much a measuring instrument, operating under specified operating conditions, will be down for routine maintenance. One way of looking at this is the expected mean time between scheduled maintenance actions. Selecting measuring instruments that have longer times between scheduled maintenance assures more uptime and may be more cost efficient.

SETUP TO MAKE THE MEASUREMENT

The setup of a measuring process is critical. This is where the article to be measured, the measuring system, and the equipment are orchestrated to provide an adequate measuring process. Figure 8-3 illustrates this relationship. A procedure or work instruction should specify the requirements to be met during setup.

It is important to the integrity of the measurement process that:
- the dimension and datum are clearly specified on the drawing so the proper measurement process is planned;
- the article is set up for measurement using locators on the same datums used for making the article;

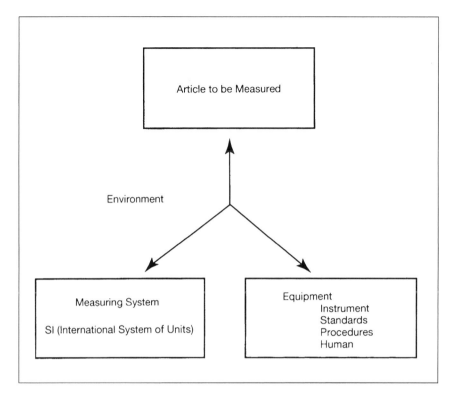

Figure 8-3. The measuring process consists of the article to be measured, a measuring system, and closely monitored measuring equipment. Also critical is the environment.

- the standard for comparison is compatible with the intent of the measurement and the instrument used; and
- the instrument can measure the dimension from the datum.

Environmental Problems

The environment is critical to the measuring process. Absolute stability in the environment, although desirable, is seldom possible, particularly in manufacturing settings. Temperature variations have a well-known effect on dimensions, as well as the shape of parts and assemblies. Variations caused by temperature also adversely affect gages and measuring instruments. Reasonable precautions should be taken to avoid extreme or sudden variations in temperature.

Because of these problems with the environment, measurements should be made in special areas that are air conditioned and humidity controlled.

These special areas are often called gage or metrology laboratories. Often, vibrations are damped out in these areas by isolators. Care is taken to stabilize the temperature of the article to be measured before the measuring process begins.

For these laboratories, the temperature is typically maintained at 68° F (20° C) and humidity ranges from 35-45%. Air pressure within the area is often higher than outside to keep dirt out. Entrance to the area may be only by double doors and special pass-through windows. Lighting is carefully designed to provide a shadowless environment and sufficient illumination for precise measurements.

Most measurements in manufacturing are made either in the manufacturing process or in close proximity. Time constraints and real time usage of the measurements do not permit use of a gage or metrology laboratory. Examples include measurements to drive process control devices.

The moves to greater accuracy, accompanied by finer positioning requirements, have greatly increased the number of gaging and inspection applications at the point of manufacture or in its close proximity. Higher accuracies usually require feedback devices in the manufacturing process for driving process control. Typically, the expanded applications consist of:

- measuring machines and comparators;
- high accuracy machine tools;
- diamond lathes for optical parts;
- facing lathes for magnetic storage disks used in computers;
- grinding machines for ferrite parts; and
- high precision equipment used in material handling.

Obviously, the environment at the point of manufacture or in close proximity will not be controlled to the extent ideally wanted. Although a building may be air conditioned, material being processed still may undergo sizeable temperature changes through various stages of completion. For these applications, adjustments should bring the measurements in line with what would be realized in an environmentally controlled area.

CAUSES OF MEASUREMENT ERRORS

Errors do occur in measurement and are abundant for most companies. One guideline advises that the first step in any corrective action effort is to recheck the measurement. An error in measurement can cause wrong conclusions that lead to decisions that delay or prevent a permanent solution to a pressing problem. Obviously, errors in measurement are critical and really cannot be tolerated.

Errors also cannot be tolerated in judging whether a product is okay. Figure 8-4 shows the causes of measurement errors that lead to wrongly rejected shipments from suppliers. The major cause is that the gages lack capability. Specifically, the variation in the gages consume over 30% of the total variation. To be capable, it is commonly agreed that the variation in a gage should be less than 10% of the total variation, but certainly not greater than 30%. Using the wrong measurement method and wrong equipment are also major causes of errors in judging parts from suppliers.

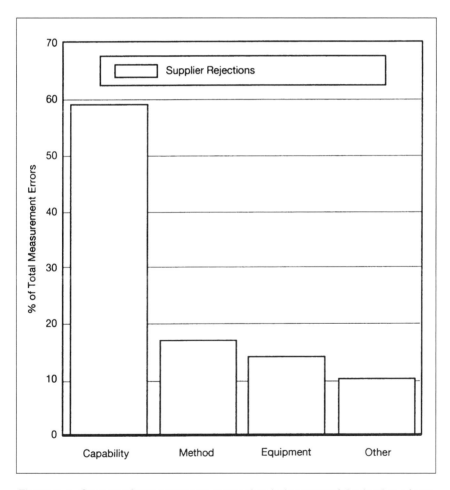

Figure 8-4. Causes of measurement errors that led to wrongful rejection of supplier product.

From a practical standpoint, there is much interaction among the errors in some measurement processes. This complicates finding the causes of the errors. Often, statistical studies, such as design of experiments, are necessary to sort this out.

Causes of errors can be grouped into three categories, with each category referring to the type of error produced.

Systematic or Fixed Errors

A systematic error remains constant or varies in a predictable way and is an unwanted offset in the measurement process. Because of its regular occurrence, it can be detected through measuring a known dimension. However, its regular occurrence is a two-edged sword since it appears legitimate and is often overlooked.

The measurement will vary from the known dimension by a constant or otherwise predictable amount. Biasing is analogous. There are several causes of systematic errors.

Calibration

Incorrect calibration occurs because of a discrepancy between the input signal and the instrument reading. It could be due to such things as a probe being too long.

Human Error

Different people usually read instruments from unlike angles. Parallax creeps in and every operator could see a different reading for the same measurement. A program of repetitive training would help ensure consistency.

Method

An incorrect method could result in a measurement offset. This may be corrected by adopting an improved approach for making the measurements. Often it may take much trial and error before perfecting a method. When the new method is fully developed, it should be documented for future guidance and consistent use.

Random or Chance Errors

A random error reflects the variation when repeated measurements are taken. It could be visualized as the spread of a normal distribution. Random errors are often recognized by the lack of consistency in the data collected. Since random error happens all the time and is often quite noticeable, it is relatively easy to find. It is analogous to precision, discussed in the previous chapter.

Judgment

Different people make different judgments. In fact, the same person may not act the same throughout the day because of personal problems and other factors. All of this contributes to variation that is somewhat random in nature. Training may help in bringing a consistent focus to the job.

Instruments

Variation may be due to the instrument needing repair. Perhaps a soldered connection may be loose. Sometimes this is best found by seeing if a redundant instrument can solve the problem.

Procedure

Often this error occurs because too much is sought from a measurement process. For example, one measurement to get the thickness of a plate may be not enough. If it is not flat or parallel, more measurements are necessary. This can be corrected by an appropriate method that is documented and followed consistently.

Unacceptable Errors

An unacceptable error is one that cannot be condoned or excused under any circumstances. It just should not have happened. Unfortunately, there are many errors of this type committed by all companies.

Mistakes

Anyone can and does commit mistakes—from the operator to the person doing the calibration to the team that selected the measuring process. These mistakes may have dire consequences on the integrity of the measuring system. A conscious positive effort should be made to reduce and stamp out mistakes. Understanding why mistakes happened is the first step. The second step is to plan an approach so that the same mistake does not happen again. This procedure should be documented and used consistently.

Computational

Errors in computation can happen anytime calculations are made. An error can happen while an instrument is set up or while a measurement is converted. Sometimes the errors are undetected if they are not obvious or do not affect a manufacturing process. With so much depending on correct measurements, it certainly would be reasonable to double check calculations to make sure they are right.

Chaotic

Chaotic errors are extreme in nature and caused by major disturbances. Causes may be vibration, shock, noise, water leakage, and the like. They are obvious and the short-term solution is to stop measuring when they present themselves. Plans should be prepared to prevent these disturbances from reaching the point where measurements are made.

Bibliography

1995 Grolier Multimedia Encyclopedia. Windows/MPC Version 7.01, 1995.

Busch, Ted, *Fundamentals of Dimensional Measurement.* Albany, NY: Delmar, 1989.

Farago, Francis T. and Curtis, Mark A., *Handbook of Dimensional Measurement, 3rd ed.* New York: Industrial Press, 1994.

Gimmi, Kenneth J., "Measurement Uncertainty." Milwaukee, WI: 1993— ASQC *Quality Congress Transactions.*

Grove, John W., editor, *Handbook of Industrial Metrology.* Englewood Cliffs, NJ: Prentice Hall, 1967.

Harral, William M., Overview of ISO 10012-1:1992 "Quality Assurance Requirements for Measuring Equipment." Dearborn, MI: *SME GAGETECH,* 1993.

Juran, J.M., editor, *Quality Control Handbook, 4th ed.* New York: McGraw Hill, 1988.

Manolis, Steve A., "Gaging, Inspection, Resolution, and Accuracy." *Quality,* May, 1993, p. 27.

Meadows, James D., "Measurement Science vs. Measurement Fact." Dearborn, MI: SME *GAGETECH 1993.*

Measuring and Gaging in the Machine Shop. Ft. Washington: National Tooling and Machine Association, 1981.

Morris, Alan S., *Measurement and Calibration for Quality Assurance.* Hertfordshire: Prentice Hall, 1991.

Prond, Paul and Ermer, Donald S., "A Geometric Analysis of Measuring System Variations." Milwaukee, WI: 1993—ASQC *Quality Congress Transactions.*

"Quality Assurance Requirements for Measuring Equipment—Part 1: Metrological Confirmation System for Measuring Equipment." International Organization for Standardization: ISO-10012-1:1992-01-15.

"Quality Assurance Requirements for Measuring Equipment—Part 2: Control of Measurement Processes." International Organization for Standardization: ISO/CD-10012-2:1993.

Salvendy, Gavriel, editor, *Handbook of Industrial Engineering*. New York: John Wiley, 1992.

Schumacher, Rolf B.F., "1001 Questions You Never Asked About Metrology." Milwaukee, WI: 1989—ASQC *Quality Congress Transactions*.

Suntag, Charles, *Inspection and Inspection Management*. Milwaukee, WI: ASQC Quality Press, 1993.

Utpal, Roy and Xuzeng, Zhang, "Relating CMM Data to Design Specification." Milwaukee, WI: 1992—ASQC *Quality Congress Transactions*.

Wick, Charles, editor-in-chief, and Veilleux, Raymond F., staff editor, *Tool and Manufacturing Engineers Handbook, Volume 4, 4th ed.* Dearborn, MI: Society of Manufacturing Engineers, 1987.

⑨

Measurement Assurance

One definition of measurement assurance describes it as a system of planned activities designed to establish and show on a continuing basis that the uncertainty of each measurement is suitably small compared with its intended use. One could easily infer from the word "establish" that measurement assurance starts early in the planning phase. The phrase "show on a continuing basis" suggests ongoing activities for controlling the measurement process while production takes place.

DIMENSIONAL ASSURANCE

To improve planning for measurements during product development, some companies in the automotive industry have adopted a formal effort, which could be called dimensional assurance (Figure 9-1). While not substituting for quality planning during product development, dimensional assurance is superimposed as part of it, increasing the likelihood that a good measurement system will be developed. Dimensional assurance principles are general enough to be adopted as a separate effort or may be incorporated into the quality planning activities of the product development cycle. The principles are fundamental and make good sense for any company, whatever the size. Dimensional assurance, which has its roots in the U.S. car companies' efforts to improve quality and stave off foreign competition, is a planning process consisting of the following steps.

Defining Dimensional Objectives. Key dimensions that will influence the perceptions of customers about quality are identified. For a vehicle, customer perceived quality items may be such things as door to fender gaps, and quarter panel to deck lid flushness. Other dimensions that have a high priority on a seriousness classification of characteristics also could be included for discussion at this stage.

A meeting is held between team members from involved areas such as styling, product design, production engineering, and the plants. Sometimes

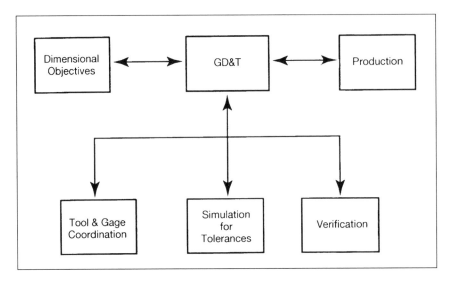

Figure 9-1. Planning process for dimensional assurance.

data is collected from clinics to help decide important issues about the point at which quality is perceived to be acceptable. The clinics normally consist of potential customers. The deliverable from this step is that buy-in is reached by the various disciplines on the quality and dimensions needed to meet customer expectations. This step is completed early in the planning stage before product designs are committed and are still easy to change.

Coordination of Tooling and Gaging. This step concerns getting buy-in from the various disciplines on the datums to be used consistently for locating and holding on the product design, tooling, and gages. It is important that the locating schemes have sound geometry when the product is designed.

Inconsistent use of datums among the plants was a major cause of dimensional variation in the past. Many dimensional problems are now avoided by parts being located and held the same way at all facilities during fabrication, gaging, and assembly.

For gages, it is critical that a *functional* approach is adopted, meaning the gage must check a part when that part is installed. If the part fits the gage, it will install properly at assembly. The deliverable of this step is proper coordination to assure consistency of the datums used during the entire manufacturing and assembly processes. This coordination must start in product development and continue during the design and build of tool-

120

ing and gages. If many parties are involved in the design and build, this effort will be very challenging. Nevertheless, it is absolutely essential to assure that dimensional objectives are achieved.

Adoption of Geometric Dimensioning and Tolerancing. Geometric dimensioning and tolerancing (GD&T) is used to specify datums and tolerances on the product design. Often, this is an iterative process with tolerances found by computer simulation. Once completed, this step delivers designs that are clearly dimensioned.

Statistical Simulation for Tolerancing. Statistical computer simulation is often used to find out what tolerances are needed to achieve dimensional objectives. Tolerances are the amount of variation that can occur without affecting the part's intended function. Based on the assembly, tolerances are initially allocated to the components and then to the parts.

Using the initial tolerances as a starting point, results are reviewed by a team of relevant parties from suppliers and within the company. Realistic tolerances in each situation are discussed. Process capability information is reviewed and assessed as to the effect on dimensional objectives. This analysis is done using a combination of common sense, process capability studies, linear statistical analysis, and 3D simulation methods.

The results of these studies are given to those concerned—engineers, part sources, and the assembly plant. Where measurement objectives are in danger, alternate approaches are evaluated, which might include different fabrication processes, assembly processes, and product designs. An iterative approach is used until all tolerance problems are resolved.

Assembly drawings, as well as component and part drawings, are then made with the agreed tolerances for each stage in the process. GD&T is used as the communication medium. A comprehensive road map of variation then exists, defining the tolerances required to meet the dimensional objectives.

The deliverable of this step is that buy-in is reached as to appropriate tolerances on the product design. GD&T dimensions are adjusted to reflect the agreements, which play a critical role in selecting dimensions and tolerances needed in the tools and gages to build the product.

Verifying Simulation Results. Checks are made to assure that all tools conform with the proper locating schemes. Gages are checked to see that they conform and that all operations are going according to plan. Also checked are tolerances on tools and gages that are usually much tighter than part tolerances—generally 10% of the part tolerance.

Parts and assemblies at each stage of development and production are closely checked to assure that they meet dimensional objectives. In this way, the dimensional assurance effort is evaluated. Necessary adjustments are made, despite being potentially costly at this time. Build problems are identified and solved by a combination of such things as common sense, statistical studies, and 3D computer simulation results.

Support for Production

Dimension assurance continues through production to solve problems in meeting dimensional objectives. Although some problems may remain, results show that the vast majority of problems are successfully dealt with by the time production starts. During the recent launch of the Neon® automobile, it was reported that 99% of measurements taken met simulation expectations using this approach.

GAGE OR MEASURING SYSTEM CAPABILITY

The precision/tolerance (P/T) ratio has been used for years as a benchmark for gage or measuring system capability and is still widely used today. However, in the past decade, statisticians have suggested an improved ratio called gage repeatability and reproducibility (GRR) that is more conceptually correct. This new benchmark for gage or measuring system capability is being widely adopted, especially in the automotive industry.

Precision/Tolerance Ratio (P/T)

For a long time, the P/T ratio has been important in evaluating how good a gage or measuring process will be in making measurements. P/T furnishes critical input as to the gage or measuring capability, relating the precision of a gage or measurement process to the product tolerance. The formula for P/T is:

Precision of Gage   P/T = Precision of a gage/tolerance of part

Historically, a ratio of 1:10 maximum has been sought for P/T by gage makers. Yet, since this ratio has been frequently exceeded, some feel that the gage is okay to use if P/T can be held below 3:10. Determining if the needed P/T is arbitrary depends on the judgment of the parties making the call and on what can be made within practical economic boundaries for a particular approach. If the P/T ratio is too large, alternate methods of measurement should be sought. One could say that a smaller P/T ratio leads to a better measurement system. Still, one has to consider how much it costs to get this added precision in the measuring system. There is, of course, a point of diminishing returns.

For each new measurement application, the P/T ratio should be checked for an ineffective measurement process. Remember, however, there may be effort and costs expended for seeking unneeded high levels of perfection. Selecting the right measurement process for each job can be helped by looking at the P/T ratio.

Gage Repeatability and Reproducibility (GRR)

Instead of P/T, a ratio called gage repeatability and reproducibility (GRR) is being adopted by many companies to assess gage or measuring system capability. There are two differences.

First, total variation (TV) replaces tolerance in the denominator. TV includes variation due to the equipment, operator, and part. Second, variation due to the equipment and operator called R&R is used in the denominator. R&R is analogous to precision. The formula for GRR is:

$$GRR = R\&R/TV$$

ASSURING STABILITY AND PRECISION

Assuring both the stability and precision of a measurement process is important. Stability is the ability of the measuring process to remain sound over time. Likewise, precision is the spread of measurements handled by the measurement process. Obtaining valid measurements is not possible unless both happen as expected.

Measurement Process

The broad acceptance by the manufacturing community that measurement is a process recently led to a new approach that allows measurement to be monitored and controlled like other processes. Control charts, long used in manufacturing, can monitor measurements on a real-time basis. Sudden or gradual deterioration of the quality in measurements can be detected. The deterioration could be from either an increase in the random error or the systematic error, or both.

By this approach, an estimate is obtained as to the value of the bias and precision of a measurement process. Bias values can be used for calibration. However, precision values could be extremely useful during troubleshooting.

Variation in a Measurement Process

With measurement being a process, variation in measurements can be explained in terms similar to that used for other processes.

Within Operator Variation. This is variation from a scatter of readings when measuring is done by the same operator, with the same measuring system, on the same product. Often, this is what is tracked when using control charts to monitor measuring processes.

Between Operator Variation. This is variation between a scatter of readings taken by two operators using the same measuring equipment and product. Often, this is due to differences in techniques used by each operator. Between operator variation is tracked by control charts used to monitor measuring processes.

Materials Variation. This variation is due to the material itself, and has nothing to do with the operator or measuring system. Unfortunately, this situation is difficult to assess since testing may change or destroy the material.

Measuring Equipment Variation

Measuring systems are subject to many sources of error, both within a single instrument and between measuring systems. Robustness of a measurement system such as sensitivity to environmental changes, could be a major problem. Fixturing could also cause variation in readings.

Test Procedure Variation. There will be variation when more than one test procedure is used to conduct measurements. Seeking the best approach, among those used, often leads to a solution.

Interlaboratory Variation. Variation among different laboratories doing the same measurements on the same product is a major problem. This problem can exist both within and between companies.

Composite Variation

The total variation from all sources is called the composite variation. Where other types of variation are independent, composite variation is the sum of those variations that made a contribution. However, in practice, there is often at least one source dependent on another source of variation. In this case, a modified approach must be used to find the composite variation.

Standard Procedures

Development of standard procedures and systems for monitoring measurements often is required to reduce the variation problem. For tracking the measurement process, check references and control charts are used. Check references are items that have known dimensions. Control charts, using the check references, can then help point out where stability and precision are out of control and must be corrected.

Check Reference

To monitor both within operator and between operator variation, a stable item needs to be measured over a long period of time. A check reference can make a good estimate of the total variability of the measurement process.

A check reference is usually a part similar to the part being measured. The check reference is measured by another process that is more accurate and this measurement should be recorded. Having a measurement traceable to a reference standard would be desirable. By using a check reference, one gets an indication of the drift and variability in the measuring process.

X-bar Statistical Process Control Chart

When the average measurement of a check reference sample goes beyond the control limits, one must strongly suspect that the measuring process is no longer stable and has drift. The apparent bias must be checked through calibration and the measuring process reset. Figure 9-2 shows stability being assessed with an X-bar chart using check references.

Range Statistical Process Control Chart

When the range of a check reference sample goes beyond the control limits, the precision of the measuring process no longer may be adequate. The cause of this apparent loss of precision must be identified for corrective action. Figure 9-3 shows the precision of a measuring system evaluated by a range chart that plots data from measuring a check reference.

Control Limits

Control chart limits must be set using a check reference and data from a measuring process that is in statistical control and stable. The process also must have a level of precision that is acceptable, with no bias. These control limits will be used as the standard to judge the performance of the measurement process.

There could be two possible reasons—or a combination of both—for using statistical loss control when measuring against the check reference. The first is that the measuring process has changed and needs correction;

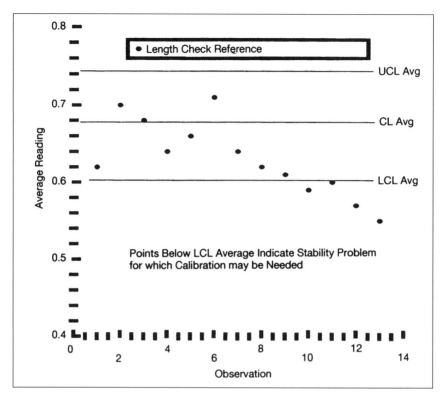

Figure 9-2. X-bar chart showing drift in average reading of check reference.

the second is that the check standard has changed. The measurement process should be stopped until the reason is found for the variation and statistical control is restored.

Samples

In general, the same approach described in any statistical process control book is used for setting control chart limits, taking data, and plotting the control charts. However, there are some things that apply to measurement processes that need clarification. For example, the sample size for measuring the check reference is usually kept small because of the time constraints. Usually five pieces are in a sample.

Regarding timing of the samples, the most demanding procedure checks the reference standards every time a batch of like items are to be measured. But samplings taken on a less frequent basis may be acceptable. Typically,

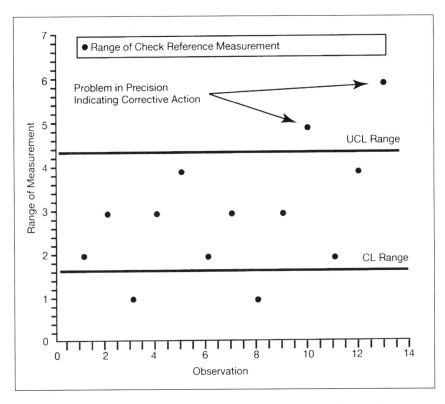

Figure 9-3. Range SPC chart showing change in precision of measuring system on check reference.

it would be more frequent for a new measuring process. On the other hand, sampling during old measuring processes with a history of being in control may be much less frequent. The frequency depends on a qualitative assessment of how well the measurement process has remained in control in the past, how much assurance is desired regarding the measurement process, and how critical the results of the measurement process are to the product.

Measuring check references at regular intervals may hide significant problems. If always taken at the same time of day, measuring system performance on the opposite shift may not be known. Examination of check references should be performed randomly.

Two techniques are used to compare the sample to the check reference. The first is to take multiple measurements using the same check reference until the sample size is achieved. The second technique is to have the same number of check references as the sample size. It is important that the measurements on the check reference are done in exactly the same manner as on the actual product. Otherwise, data from the check reference may not be representative of the actual measurements made on a product. Variability of the measuring process is indicated when the check reference is compared to the sample. Changes in variability are also obvious.

Bibliography

Gimmi, Kenneth J., "Measurement Uncertainty." Milwaukee, WI: *ASQC Quality Congress Transactions*, 1993.

Grove, John W., editor, *Handbook of Industrial Metrology*. Englewood Cliffs, NJ: Prentice Hall, 1967.

Juran, J.M., editor, *Quality Control Handbook, 4th ed.* New York: McGraw Hill, 1988.

Juran, J.M. and Gryna, Frank M. Jr., *Quality Planning and Analysis, 3rd ed.* New York: McGraw Hill, 1993.

Larsen, Curt and Brown, Don, "Achieving Dimensional Quality and Reducing Costs: A Case Study at Dodge/Plymouth Neon." Dearborn, MI: SME GAGETECH, 1994.

Prond, Paul and Ermer, Donald S., "A Geometric Analysis of Measuring System Variations." Milwaukee, WI: *ASQC Quality Congress Transactions*, 1993.

Quality Assurance Requirements for Measuring Equipment - Part 2: Control of Measurement Processes. International Organization for Standardization: ISO/CD-10012-2:1993.

Salvendy, Gavriel, editor, *Handbook of Industrial Engineering*. New York: John Wiley, 1992.

Schumacher, Rolf B.F., "1001 Questions You Never Asked About Metrology." Milwaukee, WI: *ASQC Quality Congress Transactions*, 1989.

Suntag, Charles, *Inspection and Inspection Management*. Milwaukee, WI: ASQC Quality Press, 1993.

Troxell, Joseph R., "Variance Components in Measurement Assurance Systems." Milwaukee, WI: *ASQC Quality Congress Transactions*, 1992.

Van Nuland, Yves, "Do You Have Doubts About the Measurement Results, Too?" *Quality Engineering*, V6 N1, p. 99.

Calibration

The calibration process aims to find the difference between actual readings of an instrument or measuring process and the appropriate standard. Another way to view calibration is that it finds the accuracy of an instrument or measuring process by measuring the bias.

Calibration should be conducted periodically under a specified set of operating conditions. For example, temperature may need to be controlled within a certain range. Usually a special room, called a gage or metrology laboratory, is used for calibration to assure that the specified set of operating conditions is met. If these conditions are not met, then the results of the calibration may not be reliable.

Data collected during calibration is used to find the correction needed. The correction is a value added to that obtained by the measurement process to compensate for any bias. An adjustment may be made to the instrument or measurement process to make the required correction or the correction may be added manually to the value found when measuring. In either case, the correction acts to adjust for the bias that exists in the instrument or measuring process.

The remainder of this chapter discusses three items very important to the integrity of calibration. They are the traceability of the reference standard, the actual calibration process, and a confirmation system to help assure that calibrations are done when needed.

TRACEABILITY OF REFERENCE STANDARD

In calibration, traceability refers to the ability to trace the result of a measurement to a single source that is a national or, in reality, an international standard. All industrialized countries have a national bureau of standards that constructs and maintains their national reference standards. In the U.S., this is the National Institute for Standards and Technology (NIST). The national bureaus of standards in all well-developed countries maintain ties with each other through the CGPM and agree on definitions of international standards. In this way, there can be reasonable assurance that products made in various countries are compatible.

As shown in Figure 10-1, the tracing must be by an unbroken chain of comparisons, often called a traceability chain. Every instrument in the chain is calibrated against a more accurate instrument immediately above it in the chain. What is illustrated is hypothetical. The actual traceability chain for a company will likely be more complex, involve more steps, and have a less direct path.

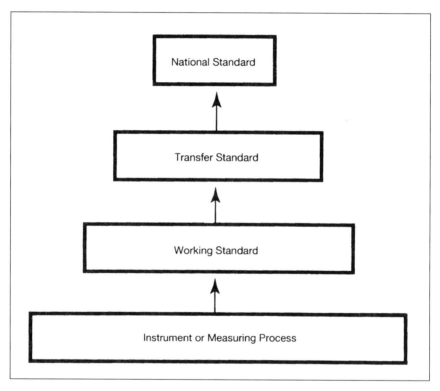

Figure 10-1. Unbroken chain—often called a traceability chain—to a national standard.

An instrument or measuring process is calibrated, possibly in the gage crib of a company, with what may be called a working standard. Conversely, a company may use an outside calibration laboratory, which would have the working standard. Of course, the working standard has a higher degree of accuracy than the instrument or measuring process, often 10 times as much.

Transfer Standards

The working standard is checked on a regular basis, against what may be called a transfer standard, which could have four times the accuracy of the working standard. Having the 10-times ideal ratio may not be practical. This is the case when there are several laboratories and working standards involved in the chain.

Few companies have their own transfer standards and the capability to make comparisons. Most use outside laboratories for transfer standards. There are many independent laboratories involved in calibration. As the traceability chain goes toward the national standards, the laboratories become more sophisticated. In more sophisticated laboratories, interferometers may be used to compare master gage blocks very accurately. In other laboratories, light waves might take the measurement of a working standard directly.

Transfer standards may be sent at specified intervals to national agencies like NIST for comparison to the national standards. Often, this is done on a sampling basis because of the substantial numbers involved. The frequency is often based on the stability and wear resistance of the transfer standard. Although documentation is not normally required, the national standards of a country are directly traceable to the international standards defined by CGPM.

With the dramatic increase in precision measurement, a single national laboratory cannot do all the calibrations required. This has led to the creation of many well-equipped calibration laboratories, within large companies and as independent businesses. Close rapport is maintained with the national laboratory so that the results of calibrations are uniform. Nevertheless, more work in this area is necessary. As mentioned in the last chapter, reducing variation among the many laboratories, both inside and outside of companies, is a major challenge in the U.S.

When a transfer standard is not an artifact, a highly sophisticated laboratory can compare it with a national standard without going to NIST. For example, the meter is now defined as the speed of light in a vacuum and a measurement can be made if the right equipment is available. However, only a few independent laboratories can justify the purchase and maintenance of such sophisticated equipment.

If all measurements have a proper traceability chain, they are consistent with each other no matter who makes the measurement. This situation allows parts to fit together in an assembly where suppliers are from various parts of the world. Military contracts and some commercial contracts specify that the calibration of measuring and test equipment be certified as traceable to national standards at NIST. International standards concerning quality systems also require traceability.

A laboratory performing calibration services normally issues a certificate of calibration. This document has the:

- specific and unique identification of the item calibrated;
- date the service was done;
- accuracy that was found;
- conditions under which results were obtained; and
- confirmation of the traceability of the working standards used in the calibration to national standards.

In the traceability chain, the most skilled metrologists are involved at the top. The technicians at the bottom of the chain are not as skilled. Only a few are involved at the top, while there are literally millions at the bottom. Training of those at the lowest level is constantly a challenge. So, too, is the design of measuring processes that are rugged, user friendly, and that can withstand the rigors of a production environment.

CALIBRATION PROCESS

A view of calibration was illustrated back in Figure 7-2. Basically, it is measuring the bias of an instrument or measuring process. The value of the bias can be used to make a correction.

In doing the calibration, the error due to the calibration process itself should be a maximum of 1:10 of the permissible error of the instrument or measuring process being calibrated. If the error in the calibration process is higher than this, the value obtained for the bias of the instrument or measuring process may not be usable for a correction.

Sometimes, the instrument or measuring system can be adjusted to account for the bias. Other applications may require the correction to be added to the measurement. If the bias is too great, the equipment may have to be repaired or thrown away. Some people feel that the instrument can be used for less critical measurements instead of being thrown away. However, this may be a mistake. Having the instrument around increases the risk that it will be used again inadvertently for the same application.

There are some devices requiring a periodic check quite different from calibration. An example is a probe used to check whether a hole was punched in a die or drilled in a machine. In such cases, only a functional check is needed to assure that the probe is still working. If not, the probe just needs to be fixed. Calibration is not pertinent.

Proper Procedures

It is important that adequate procedures and work instructions be provided for making calibrations. Consistency in using the proper method can only be assured through written instructions. How simple instruments, like micrometers, are calibrated is found in many reference books. Other information may be available from the equipment manufacturer. Technical societies, like the American Society for Testing and Materials (ASTM), are also potential sources of proper calibration methods.

A judgment should be made about whether just one instrument or the entire measuring system must be calibrated. This choice could depend on how much integration and interaction are among different parts of the measuring setup.

Of course, the setup and control of environment, as discussed in Chapter 7, are critical for calibration. However, particularly in manufacturing, the actual operating conditions may be quite different from the reference conditions specified for calibration efforts. Doing the calibration under actual operating conditions may be desirable. If this is not done, allowances should be made for the differences between operating and reference conditions. An example of this would be a difference in temperature where the thermal expansion is different. The reference temperature may be 68° F (20° C), while the actual operating temperature might be 88° F (31° C).

Frequency of Checks

The frequency of calibration checks is very important. If they are not frequent enough, the measuring process will be off. If they are too frequent, costs will be too high. There are really two basic criteria that must be balanced, the risk of measuring equipment not conforming to specification, and the costs of doing the calibrations and maintenance. Ideally, both the risk and the costs are kept to a minimum.

New equipment and setups should be calibrated before they are used. Sometimes, an instrument manufacturer's claim regarding accuracy needs to be modified. Differences between the claim and actual performance could be due to such things as unique operating conditions at the company doing the calibration. In other cases, a suspicion may exist that the measuring process may not be stable because of arduous operating conditions. Some may tighten the performance limits from those stated by the manufacturer. Others may decrease the time between calibrations. If equipment does not meet performance standards, it may have to be returned for replacement.

A sense of calibration frequency may be lacking when there is no experience with the equipment. A word of caution is in order. Frequencies used by other companies may not be right. Conditions may be very different. However, much insight can be gained by looking at:

- recommendation of the equipment manufacturer;
- criticality of measurement if wrongly accepted;
- contract requirements;
- extent and severity of use;
- influence of the environment;
- accuracy of measurement sought;
- state-of-the-art of part being measured;
- state-of-the-art of measuring process;
- sensitivity of measuring instruments; and
- parts being checked.

If a variety of state-of-the-art parts is to be measured using state-of-the-art measuring equipment, it may be that a calibration is in order each time a new batch of parts is measured. At the other extreme, if stable conditions are expected, calibration can normally take place less frequently. The frequency of calibrating existing setups should be evaluated periodically to make sure changes in conditions are properly considered. Factors influencing the analysis are:

- out of control conditions with measurement assurance;
- type of equipment;
- recommendation of the manufacturer;
- trends found during past calibrations;
- history of maintenance required;
- extent and severity of use;
- tendencies to wear and drift;
- sensitivity of measuring instruments;
- frequency of cross-checks with other measuring equipment;
- environmental conditions such as temperature, humidity, and vibration;
- accuracy of measurement sought;
- criticality of measurement if wrongly accepted;
- state-of-the-art of part being measured;
- state-of-the-art of measuring process;
- parts being checked; and
- contract requirements.

With this many factors, it is very possible that a change has taken place, making a different frequency for calibration more appropriate. For example, if history shows little bias during calibrations, the time between calibrations might be made longer. Conversely, if the measurement process is out of control between calibrations, the length of time may be shortened.

CONFIRMATION SYSTEM

A formal program is essential to assure continued accuracy of instruments or measuring processes throughout the production life cycle. Many call this a confirmation system. Such a system is aimed at preventing measuring errors beyond what is normally expected.

Acceptable Limits

A confirmation system detects errors and takes prompt action to correct them. In other words, the purpose of a confirmation system is to assure that the errors in measurement stay within acceptable limits. To accomplish this, each instrument or measuring process is checked periodically, according to a schedule. There are factors important to any confirmation effort that should be considered.

Responsibility and Authority

The functions within a company responsible for specific calibration and maintenance efforts should be clearly identified. Detailed responsibilities within various functions also need to be clarified. Who does the planning, record keeping, and other duties? Particularly important is defining who has the authority to remove an instrument or measuring equipment from production for repair and calibration. Expectations for timing also should be documented.

Labeling and Calibration Records

When a new instrument or measuring process is received, it is assigned a unique serial number and entered into the calibration program. This number is normally engraved on the measuring device to make it as permanent as possible. For smaller instruments, the marking is often placed on the case holding the device. Labels made of paper are avoided since they can be easily damaged or removed during use, making further identification difficult. Most challenging is assuring that devices that are part of production equipment, tools, and dies, also are included. To enter the new instrument or measuring process in the calibration program, a calibration record is prepared. Some companies still enter each device on a separate card for filing. However, many companies now use computer software.

Inspection and Measurement

Inspection
All measuring equipment is inspected before use. This includes verification of calibration records supplied by independent laboratories and other sources. For new or repaired equipment, calibration may be done to assure that it meets specifications. If it does not pass inspection, a rejection tag is attached and the equipment is returned to the supplier.

Scheduling
A schedule for calibration of instruments and measuring processes is prepared regularly, based on information in the calibration record. For many companies, whoever has the instrument or measuring process is notified that the device must be returned to the calibration laboratory. This poses typical scheduling problems that must be overcome to avoid shutting production down. Sometimes work must be done at odd times and very quickly. Other times, a company keeps duplicate instruments that can be used while calibration takes place.

Doing calibration on time is usually a challenge. When the instrument is not returned by the specified time, a visit to the calibration location is often necessary to pick it up. To avoid scheduling problems, some companies do calibrations where the instrument or measuring process is located. This may be absolutely necessary when a measuring process is part of a manufacturing process, die, or tool. Carts are used that have the necessary equipment and working standards for the calibrations. In some companies, a color-coded tag attached to the instrument shows when a device needs calibration. For example, if something needs calibration in January 1997, a small yellow tag will give that information.

Other schemes may be used to indicate when something must be calibrated. For example, companies may install elapsed time meters. These meters record the number of hours that a specific piece of equipment operates. By this method, the calibration frequency can be based on actual usage of the equipment. In any case, when calibration is completed, results are entered on the calibration record.

Documentation
A calibration record documents the information on each instrument or measuring process. It usually contains items such as the serial number, description, storage location, usage location, calibration frequency, acceptance criteria, special instructions, calibration date, identification of gage technician, working standard, date of next calibration, calibration results, results of measurement assurance, and products checked by the equipment.

Storage and Handling

Those handling the instruments and measuring processes must protect against damage due to handling or environmental conditions. This damage could occur during transporting, storing, or usage. If the equipment appears damaged, it must be considered nonconforming. A hold tag must be applied until it is checked out and calibrated.

Suppliers

Often instruments and measuring processes are provided to suppliers by companies that purchase their parts. Normally, the instruments or measuring processes were designed and built by the company that buys from the supplier. Here, the purchase order should require that the equipment be maintained, calibrated, and controlled by procedures equivalent to that of the company that furnished the instruments or measuring processes.

Outside Laboratories

The working standards of a company may be shipped to an outside independent calibration laboratory. Here, the purchase order should require that the calibration be done using controlled procedures. It also should require that the specific calibration results be documented and provided to the company. Included in the results should be any out-of-tolerance conditions before calibration. In addition, the results must reflect traceability of the standards used by the laboratory to national standards kept by NIST and state the environmental conditions under which the calibration was done. The certificate number of the standard used in the calibration should be included in the report.

Reaction to Nonconforming Instrument or Measuring Process

During calibration, it may be determined that instruments and measuring processes are faulty. This must be promptly reported to those responsible for the quality of the products measured. Products measured since the previous calibration may be recalled and measured again. Those found not fit for use may be replaced.

Audit

An audit should be done regularly where the measuring equipment is used, stored, and calibrated. The objective of the audit would be to determine conformance to the company's written calibration procedures.

Nonconforming Instruments or Measuring Processes

Any nonconforming instrument or measuring process needs to be immediately taken out of production. Action should be taken to correct the problem and calibration should be completed before returning the repaired

instrument or measuring process to production. Where it cannot be fixed, the instrument or measuring system should be promptly discarded. Non-conformance may be suggested if the instrument or measuring process:
- has been damaged;
- has been overloaded or mishandled;
- shows any malfunction;
- has doubtful functioning;
- has not been calibrated when scheduled; or
- has a broken seal.

Personal Measuring Instruments

Some companies require workers to have their own tools and measuring instruments. Regardless of ownership, all measuring instruments could affect the integrity of the measurement processes. It is essential that all measuring instruments, including those personally owned, be calibrated periodically.

Bibliography

The 1995 Grolier Multimedia Encyclopedia. Windows/MPC Version 7.01, 1995.

Barringer, H. Paul, "Cost Effective Calibration Intervals Using Weibel Analysis." Milwaukee, WI: *ASQC 49th Quality Congress Transactions*, 1995.

Bremmer, Bob, "Verify Accuracy Through Calibration." Milwaukee, WI: *Quality Progress,* March 1991, p. 108.

Brown, Bradley J., "Precision Measurement with Thermal Expansion." Milwaukee, WI: *Quality Progress*, February 1991, p. 65.

Busch, Ted, *Fundamentals of Dimensional Measurement*. Albany, NY: Delmar, 1989.

Farago, Francis T. and Curtis, Mark A., *Handbook of Dimensional Measurement, 3rd ed*. New York: Industrial Press, 1994.

Grove, John W., editor, *Handbook of Industrial Metrology*. Englewood Cliffs, NJ: Prentice Hall, 1967.

Harral, William M., "Overview of ISO 10012-1:1992 Quality Assurance Requirements for Measuring Equipment." Dearborn, MI: *SME GAGETECH*, 1993.

Huntley, Les, "An Improved Support System for Practical Metrology." Milwaukee, WI: *ASQC 49th Quality Congress Transactions*, 1995.

Juran, J.M., editor, *Quality Control Handbook, 4th ed.* New York: McGraw Hill, 1988.

"Measuring and Gaging in the Machine Shop." Ft. Washington: National Tooling and Machine Association, 1981.

Morris, Alan S., *Measurement and Calibration for Quality Assurance.* Hertfordshire, England: Prentice Hall, 1991.

"Quality Assurance Requirements for Measuring Equipment–Part 1: Metrological Confirmation System for Measuring Equipment." International Organization for Standardization: ISO-10012-1:1992-01-15.

"Quality Assurance Requirements for Measuring Equipment - Part 2: Control of Measurement Processes." International Organization for Standardization: ISO/CD-10012-2:1993.

Suntag, Charles, *Inspection and Inspection Management.* Milwaukee, WI: ASQC Quality Press, 1993.

Wick, Charles, editor-in-chief, and Veilleux, Raymond F., staff editor, *Tool and Manufacturing Engineers Handbook, Volume 4, 4th ed.* Dearborn, MI: Society of Manufacturing Engineers, 1987.

Standards for Measurement

This chapter is about standards that apply to the application of instruments and measurement processes. It should be read with Chapter 4 of this book. That chapter, although featuring standards for *inspection*, has much general material that directly concerns instruments and measurement processes.

MILITARY STANDARDS

Military standards are in a transition. With the publishing of MIL-HDBK-9000, contracts can now require compliance with ISO 9000 standards. MIL-HDBK-9000 provides guidance on the application of the ISO 9000 series quality system standards to contracts. It is likely that the ISO 9000 standards will replace the current military standards as requirements in future contracts.

Until now, MIL-Q-9858A was used solely to specify conformance to desired quality program requirements. For contracts that did not require a broad-based quality system, MIL-I-45208A was used. MIL-I-45208A specifies inspection system requirements. It is used for contracts where quality can be set only by inspections or tests. Both about measurements, MIL-STD-9858A and MIL-I-45208A refer to MIL-STD-45662 for what is required for calibration.

MIL-STD-9858A provides requirements for a quality system that must be complied with by companies that contract with the government. It has many provisions concerning instruments and measuring processes.

MIL-I-45208A covers the requirements for an inspection system that must be complied with by companies that contract with the government. Concerning instruments and measuring processes, it has provisions equivalent to these in MIL-STD-9858A.

MIL-STD-45662 concerns calibration procedures for instruments and measurement processes to control the accuracy of both measuring and test equipment and measurement standards. Again, it must be complied with to satisfy contractual requirements. The objective of this standard is to prevent measurement inaccuracies by timely detection of deficiencies and prompt corrective action. It contains requirements for initially designing and then maintaining a calibration system. Besides containing definitions, MIL-STD-45662 has requirements for:

- calibration system;
- measurement standards;
- environmental controls;
- intervals for calibration;
- calibration procedures;
- out-of-tolerance conditions;
- adequacy of the calibration system;
- calibration sources;
- records;
- calibration status;
- control of supplier calibration; and
- storage and handling.

Worth mentioning is that MIL-STD-45662 limits the uncertainty when measuring the working standard during calibration. It is limited to a maximum of 1:4 of the tolerance of any product characteristic measured. If it is greater, the equipment may not be adequate for doing the calibration.

Measuring and Testing Equipment

Gages and other measuring and testing devices must be used to assure that production conforms to requirements. These devices must be calibrated as needed using standards that are traceable to national standards. Accuracy of the measuring and testing equipment, whether in-house or at a supplier's facility, is a prime concern. The objective is to adjust, replace, or repair an item before it becomes inaccurate. For calibration requirements, MIL-STD-45662 is to be followed.

It is recognized that the ability to assess measurements may be built into jigs, fixtures, tooling masters, templates, patterns, and other production processes. The same checks for accuracy must be made on these items as on gages and other measuring and testing devices. This includes an initial assessment and periodic calibrations.

The government must be allowed to use both personnel and devices at the contractor to determine if the devices conform to the requirements of the military standard.

The government must be informed if any precision measurement needed exceeds the state-of-the-art for obtaining it.

STANDARDS BY PROFESSIONAL ORGANIZATIONS

Voluntary standards developed through professional organizations have been of great assistance to those concerned with measurements. The American Society for Testing and Materials (ASTM) and the American Society of Mechanical Engineers (ASME), for example, have been very active in this area. Many of these standards are handled through ANSI jointly with the professional society. ASTM and ASME standards, some of which concern measurements, are available directly through the professional society. Catalogs of standards are available from ASTM, ASME, or the American National Standards Institute (ANSI).

Y14.5, previously discussed in Chapter 7, was developed through ASME. Although dealing with only geometric dimensioning and tolerancing (GD&T), Y14.5 has been a big help to those concerned with measuring. In 1996, a new standard, B89.3.2, will be issued by ASME about dimensional measurement methods. This will be the first to directly address how to measure using GD&T techniques.

B89.1.12M from ASME also concerns methods for the performance evaluation of coordinate measuring machines (CMMs). This standard applies to the testing of CMMs having three linear axes perpendicular to each other. It contains methods for specifying and testing repeatability, linear displacement accuracy, and artifacts.

Q9000 QUALITY SYSTEM STANDARDS

Many companies desire to be registered in the quality system standard, Q9000. For these companies, control must be maintained over many measuring processes during the product cycle. It applies to all measuring processes used in the development, production, installation, or servicing of products.

The control effort aims to provide confidence in decisions or actions based on measurement data. Control must be over gages, instruments, sensors, special test equipment, and related test software. In addition, items that can affect the important characteristics of a product or process also must be controlled. These items include jigs, fixtures, test hardware, comparative references, and process instrumentation.

Elements of Control

Q9000 discusses elements that are the basis for the control of measurements. Among these elements is proper selection and specification of any instrument and measuring process. In doing this, it is important to consider the range, accuracy, and robustness of the device, and the environmental conditions under which it must successfully operate.

Calibrating the instrument and measuring processes before they are used will help assure that the needed accuracy has been achieved. Precision of the device also should be evaluated at this time, as well as testing the software and procedures for controlling automatic test equipment before the application.

Periodic adjustment, repair, and calibration of the instruments and measuring processes are performed to maintain required accuracy for the measuring system. Frequency will vary depending on the manufacturer's specification, the results from previous calibrations, and the extent that the instrument or measuring process was used.

Important procedures and work instructions should be documented and should include:
- unique identification of the instruments;
- frequency of calibration;
- calibration status;
- scheduled calibrations;
- handling;
- preserving;
- storage;
- adjustments;
- repairs;
- installation;
- use; and
- traceability to reference standards of known accuracy and stability.

Control Procedure

Q9000 also gives guidance regarding procedures to be followed when controlling measuring processes. The following steps are suggested:
1. Find the measurements to be made and the accuracy and precision required.
2. Select the appropriate inspection, measuring, and test equipment capable of the needed accuracy and precision.
3. Calibrate and adjust, as necessary, before using all new inspection, measuring, and test equipment that can affect product quality. This should be done with certified equipment having a known valid relationship to nationally recognized standards.

4. Repeat at prescribed intervals the calibration and adjustment of all existing inspection, measuring, and test equipment that can affect product quality. Again, this should be done with certified equipment having a known valid relationship with nationally recognized standards.

5. Determine the process for the calibration of inspection, measuring, and test equipment. This should include the consideration of equipment type, unique identification, location, frequency of checks, check method, and acceptance criteria, along with the action to be taken when results are not successful.

6. Identify all inspection, measuring, and test equipment. Labeling with a suitable indicator to show the calibration status is important.

7. Maintain calibration records for inspection, measuring, and test equipment.

8. Analyze and document the validity of previous inspection and test results when inspection, measuring, and test equipment is out of calibration.

9. Assure that the environmental conditions are suitable for the calibrations, inspections, measurements, and tests being carried out.

10. Assure that the handling, preservation, and storage of inspection, measuring, and test equipment is adequate for continued accuracy and fitness for use.

11. Safeguard inspection, measuring, and test facilities, including test hardware and test software, from adjustments that would change the desired calibration setting.

The Q9000 standards refer to ISO 10012 for more information on a confirmation system for calibration.

Corrective Action

Appropriate corrective action is necessary when processes are out of control, or when inspection, measuring, and test equipment are out of calibration. An analysis should be made to learn the effects on previously completed work. This analysis should consider the need for reworking, retesting, or scrapping the work. In addition, an investigation to find the cause of the problem is important to prevent recurrence. This investigation should include, among other things, a study of the calibration methods and frequency, effectiveness of training, and adequacy of the test equipment.

Outside Testing

To avoid costly duplication or additional investment, outside laboratories may be used for inspection, measurement, testing, or calibration. However, the same standards must be met as if the activities were done in-house.

Standards are constantly evolving and being made better to provide improved guidance to those involved. Many are regularly updated by peers who provide new levels of expectation.

Bibliography

Clements, Richard Barrett, *Understanding the 1994 Revisions to ISO 9000.* Grand Rapids, MI: National ISO 9000 Support Group, 1994.

Durand, Ian G.; Marquardt, Donald W.; Peach, Robert W.; and Pyle, James C., "Updating the ISO 9000 Quality Standards: Responding to Market Place Needs." Milwaukee, WI: *Quality Progress*, July 1993, p. 23.

Guidance on the Application of ISO 9000/ASQC Q90 Series Quality System Standards. Department of Defense/NASA: MIL-HDBK- 9000/ NASA-HDBK-9000, 1994.

Juran, J.M., editor, *Quality Control Handbook, 4th ed.* New York: McGraw Hill, 1988.

Military Specification Inspection System Requirements. Department of Defense MIL-I-45208A, 1981.

Military Specification Quality Program Requirements. Department of Defense MIL-Q-9858A, 1962.

Military Standard Calibration System Requirements. Department of Defense MIL-STD-45662A, 1980.

Peach, Robert W., editor, *The ISO 9000 Handbook, 2nd ed.* Fairfax, VA: CEEM Informational Services, 1994.

Quality Management and Quality System Elements–Guidelines. Milwaukee, WI: American Society for Quality Control ANSI/ASQC Q9004- 1-1994.

Quality Management and Quality Assurance Standards–Guidelines for Selection and Use. Milwaukee, WI: American Society for Quality Control ANSI/ASQC Q9000-1-1994.

Quality Systems–Model for Quality Assurance in Design, Development, Production, Installation, and Servicing. Milwaukee, WI: American Society for Quality Control ANSI/ASQC Q9001-1994.

Quality Systems–Model for Quality Assurance in Production, Installation, and Servicing. Milwaukee, WI: American Society for Quality Control ANSI/ASQC Q9002-1994.

Quality Systems–Model for Quality Assurance in Final Inspection and Test. Milwaukee, WI: American Society for Quality Control ANSI/ASQC Q9003-1994.

Suntag, Charles, *Inspection and Inspection Management.* Milwaukee, WI: ASQC Quality Press, 1993.

Tsiakals, Joseph J., "Revision of the ISO 9000 Standards." Milwaukee, WI: *ASQC 48th Annual Quality Congress Proceedings*, 1994.

Voelker, Herb, *Constructive Consequences of the 1980's Metrology Crisis.* The Brown & Sharpe Publication of Accuracy Manufacturing, 1994.

Wadsworth, Harrison M., "Standards for Tools and Techniques." Milwaukee, WI: *ASQC 48th Annual Quality Congress Proceedings*, 1994.

Guidelines for Measurement

This chapter, about guidelines for the application of instruments and measurement processes, should be read in conjunction with Chapter 5 of this book. Although that chapter is specifically on guidelines for *inspection*, it has information also pertinent to instruments and measurement processes.

INDUSTRY-SPECIFIC DOCUMENTS

QS 9000 Quality System Requirements

In 1994, QS 9000 quality system requirements, published by the Automotive Industry Action Group (AIAG), were adopted by Chrysler, Ford, General Motors, and the major domestic truck producers. Based on the ISO 9000 quality system standards, QS 9000 has added provisions reflecting the quality movement in the automotive industry. It states that records must be maintained on the calibration and certification activity for all measuring and test equipment, including gages owned by employees. These records must include:

- revisions following engineering changes;
- gage conditions and actual readings when received for calibration;
- notification to customer if suspect product has been shipped; and
- measurement system analysis.

Suppliers must provide appropriate technical resources for gage design, fabrication, and full-dimensional inspection. For work subcontracted, a tracking and follow-up system is necessary.

Measurement Systems Analysis Reference Manual

Appropriate statistical studies must be done to analyze variation in each type of measuring and test equipment system. This applies to all systems specified in the control plan. The Measurement Systems Analysis Reference Manual should be used in performing gage repeatability and reproducibility studies (R&R).

This classic reference manual has been used as a guideline by many companies in and out of the automotive industry over the years to do R&R studies. R&R in earlier versions was a percentage of the tolerance measured. In the latest version, this was changed to a percentage of total variation, a change felt to be more conceptually sound. The goal of an R&R study is to understand and quantify the variation due to both the equipment and the operator. This is called repeatability and reproducibility, respectively—terms defined in Chapter 7.

Overall, R&R should be less than 10% of total variation. This is not always achieved and up to 30% is often allowed. The lower the ratio, the lower the random variation in the measurement. A lower ratio for a measurement process means that it has better precision. Another view is that the lower the value of the ratio, the more likely that a measurement will be dependable.

ISO GUIDELINES

ISO 10012

The two-part ISO 10012 guideline stresses that measurement should be considered a process, allowing process control methods to detect accuracy and precision problems. All measurements are considered to have variability, meaning no true measurement can be obtained. Nevertheless, variability can be reduced and controlled through control charts using check references. For accuracy, the X-bar chart is used to assess statistical control. On the other hand, precision is determined by the R chart.

Part 1 of ISO 10012 describes the main features of a confirmation system for measuring equipment that assesses compliance to a specification. Not addressed are other influences covered in the umbrella quality system ISO 9000 standards, such as methods of measurement or competence of personnel. The objective of the confirmation system described is to assure that measurements are made with the intended accuracy. It is assumed that confirmations are made when scheduled. However, the system cannot guard against random failure or if the equipment is being used incorrectly. Part 1 covers:
- definitions;
- general requirements;

- measuring system requirements;
- confirmation system requirements;
- periodic audit and review of the confirmation system;
- planning;
- uncertainty of measurement;
- documented confirmation procedures;
- records;
- nonconforming measuring equipment;
- confirmation labeling;
- interval of confirmation;
- sealing for integrity;
- use of outside products and services;
- storage and handling;
- raceability;
- cumulative effect of uncertainties;
- initial choice of confirmation intervals; and
- different methods of choosing confirmation intervals.

Part 2 concerns process control of the system setup to take the measurements. The means used may vary from control charts and check references or surveillance, to almost no control at all, depending on the application. The objective is to assure valid measurement results. Contents of this guideline are:

- documentation;
- measurement processes;
- measurement process setup and design;
- metrological confirmation system;
- system for measurement process control;
- data analysis for control of measurement processes;
- surveillance of the measurement process;
- intervals of surveillance;
- control of measurement process;
- failure of the measurement process control system;
- verification of measurement process;
- identification of verified measurement processes;
- record of measurement process control;
- personnel;
- periodic audit and review of the measurement process control system; and
- measurement uncertainties.

Bibliography

Advanced Product Quality Planning and Control Plan. Southfield, MI: Automotive Industries Action Group (AIAG), 1994.

Fundamental Statistical Process Control Reference Manual. Southfield, MI: Automotive Industries Action Group (AIAG), 1992.

Guidance on the Application of ISO 9000/ASQC Q90 Series Quality System Standards. Department of Defense/NASA: MIL-HDBK-9000/ NASA-HDBK-9000, 1994.

Juran, J.M., editor, *Quality Control Handbook, 4th ed.* New York: McGraw Hill, 1988.

Measurement System Analysis Reference Manual. Southfield, MI: Automotive Industries Action Group (AIAG), 1990.

Pince, Bruce W., et al., "Automotive Quality Requirement Standardization." Milwaukee, WI: *ASQC Quality Congress Transactions*—Nashville, 1992.

Production Part Approval Process. Southfield, MI: Automotive Industries Action Group (AIAG), 1993.

Quality Assurance Requirements for Measuring Equipment—Part 1: Metrological Confirmation System for Measuring Equipment. International Organization for Standardization: ISO-10012-1:1992-01-15.

Quality Assurance Requirements for Measuring Equipment—Part 2: Control of Measurement Processes. International Organization for Standardization: ISO/CD-10012-2:1993.

Quality Management and Quality Assurance—Vocabulary. International Organization for Standardization: (ISO/DIS) 8402, 1992.

Quality System Requirements. Southfield, MI: Automotive Industry Action Group (AIAG) QS-9000, 1994.

Tooling & Equipment Supplier Quality Assurance Guideline. Southfield, MI: Automotive Industry Action Group (AIAG), 1988.

Software for Measurement

The use of software for measurement is rapidly increasing for many reasons. One is the demand for more automatic measurements, requiring software to produce and transmit the data. Another reason is the widespread need to reduce errors in the measurement process. Some maintain that 50% of the situations where product specifications are not met are due to improper measurements. A third major reason is that most companies use the computer as a means to process and transfer information.

The result of this conversion to computers is that measurement processes are more capable than the previous manual processes. More applications are possible. More data can be kept and analysis can be made on a wider scope. Diagnostics are improved. However, the challenge is turning the great mass of data into useful information. This is the job of software.

CUSTOM SOFTWARE FOR MEASUREMENT ACTIVITIES

Custom software to support measurement activities is usually supplied by the manufacturer of the measuring process. This is the case for customized automatic inspection stations and the measurement portion of process control devices.

Customized software is also supplied by coordinate measuring machine (CMM) manufacturers. This software directs functions in the CMM, ranging from interpreting measurements to automatically driving the probe in a series of steps that measure a part. A customized program may be made up of three basic components that:

- drive the probe to the data collection points;
- measure and compare the distance traveled to the standard built into the machine; and
- translate the data into the form needed for output.

Software Types and Operation Modes

Two types of software programs are involved. The first, which does the general repetitive functions and calculations, is stored in the measurement function library. This software controls such functions as resolution selection, axis scaling, and datum selection and reset. The second type of software program, involving manipulation of the CMM, can be used to direct various general modes of operation.

There are three types of CMM operating modes.

Manual Mode

In this mode, the CMM is driven by the menu on the machine. Usually, for single-part measurements, an operator is directly involved in manipulating the machine through selections on the menu.

Automatic Mode

A program for doing the measurement task may be written by a programmer or developed during the learning mode. Depending on the type of CMM, the program may sequence the machine and drive the probe to required locations. For less sophisticated machines, the program may only coach the operator regarding the sequence to be used. A programmer is involved in writing and verifying the program. User friendliness is very important. The operator loads the program and positions the part to be measured in the correct manner. Measurements are taken and a report is printed.

Learning Mode

During learning mode, the CMM is "taught" what to do by a technician walking it through the desired sequence. Sometimes it is called the programming mode. The sequence is placed in the memory of the CMM computer and repeated for verification. Adjustments and off-line instructions are often necessary to make it run smoothly. Once these are completed, the program can be used to measure a like part.

OFF-THE-SHELF SOFTWARE FOR MEASUREMENT ACTIVITIES

Calibration Software

For calibration efforts, software can evaluate and control a company's instruments and measuring processes. The software can track such things as the location, past and future assignments, and accuracy. It can keep a history of calibrations, schedule future calibrations, and develop reports to help direct devices to the calibration area or to be shipped to an outside laboratory. Some software contains a data base of calibration procedures covering everything from gage blocks to calipers.

Data Acquisition Software

This software permits measurement data to be collected automatically. Sometimes outputs from a sensor are transmitted directly to the input of a standalone computer or a network. In other situations, instruments are connected by an umbilical cord to a portable computer. When an operator moves the instrument into a measuring position, the data is transmitted to the portable computer. Later, the data stored in the portable computer is dumped into a larger computer. In either case, the accumulated data is available for processing in statistical analysis packages.

Gage Repeatability and Reproducibility Software

Software to do gage R&R studies makes the effort much easier. These packages can be standalone or interfaced with calibration systems. Only data needs to be entered. The calculation is done by the software, eliminating the possibility of error. For many software packages, reports can be made in a format that meets the requirements of, for example, the automotive industry. Many current software packages are based on the automotive industry guideline for R&R studies.

Procedure Software

Software packages can help prepare procedures concerning measurements and calibration. They contain what is often represented to be fully developed procedures that can be printed and used. Though the text in these packages is "canned" and should not be used as is, it is stored in a format that can be changed in a word processor to reflect the different conditions in a company. Relevant procedures can be developed, using this software as a base, to meet the requirements of different system standards.

Simulation Software

Simulation software is available to help engineers predict and optimize manufacturing and assembly variation. Product and process targets and tolerances for measurements can be set. Also, process capability can be estimated while the design is still only on paper and before tooling is committed. This may avoid costly changes in the product design or manufacturing process.

Test Preparation Software

Test preparation software is available to manage and control laboratory work flow. It also can interface with laboratory instruments and communicate with manufacturing applications. Other software can provide the basis for developing test plans and managing test efforts. The user can monitor the validation of technical requirements and compute metrics. Some software can produce customized certificates of analysis from customer and product information data bases.

The American Society for Quality Control (ASQC) publishes an annual directory of software concerned with quality. Although PC applications dominate the offerings, other operating systems are listed for some programs, including DEC ALPHA, DEX VAX/VMS, UNIX, IBM mainframes, Macintosh, and others.

Bibliography

"1995 Software Directory." Milwaukee, WI: *Quality Progress*, March 1995.

Banks, Jerry, *Principles of Quality Control*. New York: John Wiley, 1989.

Busch, Ted, *Fundamentals of Dimensional Measurement*. Albany, NY: Delmar, 1989.

Farago, Francis T. and Curtis, Mark A., *Handbook of Dimensional Measurement 3rd ed.* New York: Industrial Press, 1994.

Juran, J.M., editor, *Quality Control Handbook 4th ed.* New York: McGraw Hill, 1988.

Suntag, Charles, *Inspection and Inspection Management*. Milwaukee, WI: ASQC Quality Press, 1993.

Weber, Ernest G., "Computers, Metrology and Manufacturing." Carol Stream, IL: *Quality*. January 1992.

Measuring Tools and Equipment

This chapter does not cover all measuring tools and equipment. Such coverage would be beyond the scope of this book. Rather, the chapter concentrates on measuring tools and equipment that appear to dominate plans to achieve better measurement strategies for manufacturing. These include noncontact sensors and CMMs.

This omission does not imply that the devices left out are unimportant to manufacturing. In reality, the opposite is true. Simple instruments, like micrometers and calipers, are basic and will continue to be important to manufacturing. Likewise, contact sensors are widely used by most companies. These devices, and others not mentioned in this chapter, make a valuable contribution to good manufacturing practices.

TRENDS FOR MEASURING EQUIPMENT AND PROCESSES

There has not been any recent radical measuring technology. Rather, the emphasis has been on improving what is available, from simple gages to CMMs. For example, customers are demanding that these devices be more rugged and smaller. There are, however, other trends impacting measuring technology.

Improved Accuracy and Precision

To improve accuracy and precision of measurements, a transition from single-measurement checks to multiple measurements is being widely adopted. More points are needed to check a characteristic. For example, roundness is determined by multiple measurements instead of just a single measurement to find the diameter. A better assessment of contours is also possible with multiple measurements. As a result, instruments are becoming more complex and require a computer base.

Gaging Integral with Process

More gaging is being installed as an integral part of manufacturing processes. It may be used to drive process control devices or to do inspection automatically. In addition, there are more CMMs on the manufacturing floor. Driving this change is the planning done early in the product development cycle by multidiscipline teams. With such an orchestrated approach, product quality is likely to be better.

Higher Production Output

For higher production output, parts need to be produced right the first time. To improve quality, monitoring of parts is being done more intensively both on the line and during processing. Also, the measuring process, if integrated into the production line, must often perform at a higher rate. To help, noncontact gaging is being used more because of its speed.

Fewer Workers

Since many companies are trimming their work force to maintain competitiveness, gages must perform more functions. They also must be very user friendly since operators trained to run complex equipment may be scarce. For these reasons, measuring equipment configuration is becoming very critical.

Documentation

More data handling and analysis capability built into the instruments and measuring processes allows documentation of results to be maintained in computer storage. Without this capability, meeting customer documentation requirements may not be possible.

More Flexibility

Small lot sizes cause frequent changeovers of production lines, and instruments and measuring equipment need to be flexible so they can be used on a variety of parts. Generally, machining centers or manufacturing cells are used instead of transfer lines. For measuring, modular gages and CMMs are increasingly used.

MODULAR STRATEGY

Modular measuring processes are assembled from components in an assortment of possible combinations. Selected components, often called modules, are those needed to achieve a certain measurement strategy. In a general sense, they can be viewed as a setup made with a kit of parts that can be put together in various combinations.

For example, a fixture to locate parts may be built from modules, a method often used for CNC machines. Devices for making the check could be selected from another assortment of measuring modules to meet the needs at hand. Using modular measuring processes avoids the need for dedicated devices. A setup can be made when needed, then torn down later and the modules reused in subsequent setups.

Advantages

Modular gaging allows a variety of parts to be measured. For example, a modular gage can handle a camshaft or transmission shaft despite a difference in the diameter or length. Modifications to allow checking different parts often can be done easily and may require only an adjustment or simple change of a module. Often, users can mix and match the various components of the modular gage to meet a variety of situations. The concept is very valuable when a company is faced with short lead times or where small production quantities do not call for dedicated gages. This advantage also applies when a product has an unknown life span and investment must be kept low.

When better components for measuring are available in the marketplace, they can be added to the kit of parts. Usually they can be added to an existing setup without destroying the original intent.

CHALLENGES OF TEMPERATURE

Many consider temperature the greatest cause of error when checking dimensions in a manufacturing setting. Sometimes just the heat of the operator's hand can introduce error. From a practical viewpoint, the integrity of measuring hinges on controlling temperature variations or, alternately, the ability to make valid corrections for the differences.

Designers who specify product dimensions assume the temperature of both the part and the gage is stabilized at 68° F (20° C) when checks are made. For manufacturing, unfortunately, this assumption may be true only in the gage laboratory.

Complicating this is the demand for better quality, leading to tighter tolerances on parts and assemblies. Measuring parts with tighter tolerances is a challenge. Often the capability of conventional shop floor measuring instruments is not enough. Compounding this problem is the effect of temperature on measuring results.

In most plants, there are wide temperature variations, from 50° to 90° F (10° to 32° C) or more, due to reasons not easily corrected. These may include such things as open windows, direct sunlight, or the energy dissipated from different processes. Temperature variation will affect both gages and parts. The temperature of a part could fluctuate several hundred degrees as it is being made. For example, material may have to be annealed before it can be worked.

During processing, energy from the tools will be absorbed by the material, causing additional fluctuations in temperature. When processing is complete, the part may need to be hardened by heat treating at an elevated temperature. Figure 14-1 shows how a dimension varies as the temperature changes. As shown, temperature can significantly affect the size of what is being measured. An aluminum bar exactly 4 inches (101.6 mm) long at 68° F (20° C) grows to 4.005 inches (102 mm) when its temperature is 100° F (38° C).

As the need for precision measurement grows, the need to account for even the smallest temperature variation becomes more important. Take, for example, a part that was just heated by the residual effect of friction during a centerless grinding operation. Checked shortly after coming off the machine, it was at a temperature of 73° F (23° C). The length measured 3 inches (76.2 mm) that was right on nominal. A few hours later, it was checked in the gage laboratory. The temperature of the part was stabilized at 68° F (20° C). Due to the difference in temperature, it now measured 0.0002 less. Temperature variation can drastically affect measurement results.

Small temperature changes when a part is near tolerance limits could cause the part to be wrongly rejected. Consequently, a part that just completed processing may have to be stored until it stabilizes at a lower temperature. However, stabilizing the temperature of a part is not often an option. Measurement results may be needed immediately, as is the case with in-process gaging. But, the in-process gaging may be elevated in temperature by the process or by measuring parts with elevated temperatures. For companies, it is critical that strategies are developed to account for temperature differences.

A challenge also exists when the measurement process is checking parts made of different materials. For example, aluminum expands twice as much as steel due to temperature changes. For these measuring processes, a strategy is needed to account for the different materials being measured.

Also important is the effect of temperature on a gage. Temperature differences can cause significant errors in gages despite the material used. But, the use of some material for gages causes the error to be greater. For example, many portable gages used by production workers are made of

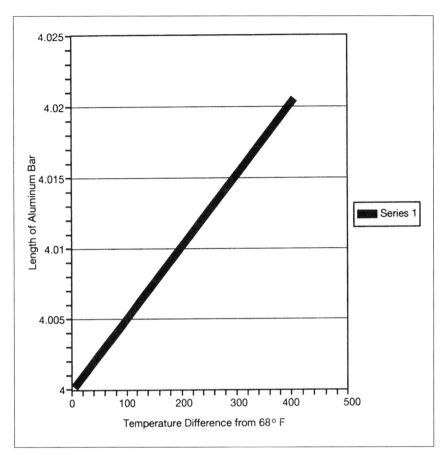

Figure 14-1. Dimensions change as temperature varies. This illustration shows how temperature changed the length of a four-inch aluminum bar.

aluminum. Because of differences in thermal expansion, the error for aluminum gages is twice that of steel gages.

Meeting Temperature Challenges

From a practical standpoint, a gage that is accurate to 0.0001 inches (0.00254 mm), or better, may be more dangerous than useful if corrections are not made for differences in temperature of the part and the gage from ambient. Wrong decisions could lead to misleading data that produces inadequate quality and disastrous results for a company.

The control for this is threefold. First, how the difference in temperature affects the structure of the gage must be determined. Second, how the part is affected by the temperature difference must be found and corrections made. Lastly, the difference in the coefficient of thermal expansion of the instrument's scale and the material in the part must be considered and adjustments made. One should consider all three factors collectively to assure the integrity of a measurement.

Temperature Variations in Gages

There are two design approaches to help account for temperature variations in gages. The first is to use materials like invar and carbon fiber composites for gages to reduce the effects of temperature as well as humidity. Although the cost is much more, the materials are very stable in spite of changes in temperature. This approach is used largely in CMMs, where a very robust device is needed and the high cost can be spread over a long period.

The second approach, which is more popular, is to use a computer system that compensates for temperature changes in the gage. It is used in CMMs on the factory floor. In these situations, the temperature of the CMM structure is about 86° F (30° C) as claimed by CMM manufacturers.

Those using the second approach make the gage out of light-weight, rapidly expanding materials such as aluminum. The aluminum reaches thermal equilibrium quickly and the heat is evenly distributed throughout the gage. With this approach, any temperature change will only have a small effect on the gage's geometry. A relatively stress-free structure is obtained. The computer system uses input from a network of temperature sensors mounted in the gage to calculate the needed corrections. Often a hood is placed over the device to buffer the system from changes in the environment.

Temperature Variation in Parts

Gages that compensate for temperature variations in parts are readily obtained in the marketplace. No manual calculations are needed to find the correction factor. These gages can display or transmit measurements within a few seconds after being adjusted for temperature.

One example of this technology is a contact sensor. When it measures a dimension, the gage also senses the temperature by using a thermocouple touching the part. A computer program uses the coefficient of expansion of the material in the part to adjust the value. The program calculates the dimension as if the part was at 68° F (20° C) and then displays or transmits the corrected value. In this way, a measurement can be directly compared with what is required, despite the actual temperature of the part.

Gages that compensate for temperature of the part can be viewed as a digital electronic indicator driven by microprocessors or computers. The microprocessor or computer provides flexibility, such as allowing gages to be customized to fit specific applications or parts. A transducer is linked to the microprocessor and a linear variable differential transformer (LVDT) in the transducer measures displacement. The measuring system uses sensors to measure the temperature of a part and the measuring device. These temperature sensors are also linked to the microprocessor.

Other Useful Features of Temperature-compensated Gages

Learning

Some devices can learn the coefficient of thermal expansion of a specific material. They do this by taking the dimensions of material at two different temperatures and then computing the coefficient. This is valuable when the coefficient is unknown.

Prediction

The dimension of a part at a given temperature can be predicted with some temperature-compensated gages. For example, it may be necessary to know the dimension of a part at its operating temperature rather than at ambient. A mold designer, for example, may wish to predict a dimension when plastic is injected at 600° F (316° C). Some gages can be set to display predicted, rather than actual, dimensions.

Computer Interfaces

Most gages have an interface allowing communication with computerized statistical process control systems. This setup also allows networking with other gages. In this way, one gage could provide a temperature correction factor for several measuring systems.

Global Temperature Correction

An effective system for temperature compensation simultaneously measures the gage temperature, the part, and the scale on the gage. After considering all three temperatures, a correction is calculated and applied to the measurement.

Though the three temperatures are used, it is noted that the corrected measurement may only closely agree with that which would be obtained later in a gage laboratory controlled at 68° F (20° C). An identical measurement is not probable because of variation induced by the measuring process and the inherent inaccuracy of the thermal expansion coefficient. However, the measurement error will be significantly reduced.

CONTACT AND NONCONTACT MEASUREMENT

Contact measurement means that either the measuring instrument itself or a probe contacts the part being measured. As with a micrometer, the contact is often accompanied by a clamping action with slight force exerted.

The term noncontact refers to any measuring process where contact is not made with the part being checked. Air gaging is one example, along with optical techniques such as laser systems and machine vision, and many other types, such as electrical capacitance sensors.

Contact Measurement

Figure 14-2 shows an exaggerated view of what happens when a probe meets the work surface. Depending on the force exerted and the shape of the probe, the surface may be permanently deformed. Playing a big influence is the material being measured. A softer material, such as copper, would deform more than steel. Penetration of the material must be accounted for when calibrating contact gaging. As an example, a stylus or pointed probe used to make surface roughness measurements could scratch the surface it is measuring.

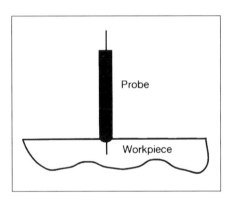

Figure 14-2. Exaggerated illustration of deformation that could happen during contact sensing.

Another challenge with contact gaging concerns the condition of the part being measured. High humidity in a plant may cause a part to rust. Dirt and other contamination may build up on the surface. The accumulated surface contamination will cause the measurements to be incorrect and may mislead those concerned. Contact sensors are felt by many to be slower than those that do not contact the part. Extra time is taken to position the probe and contact the surface. Usually only one measurement can be checked at a time. For these reasons, noncontact sensors may provide measuring systems with greater speed, precision, and accuracy.

Noncontact Measurement

Most contact sensors provide only one measurement for each touch. However, with noncontact sensors, many measurements, or their equivalent, are done in a single step. Where multiple measurements are desired, noncontact sensors are faster. Also, noncontact sensors make sense for checking fragile parts. In general, noncontact sensors deliver more accuracy and precision than do contact sensors. Thus, their role in measurement is growing.

A growing trend is to use measuring systems only on the production floor, requiring a design that assures that the systems can survive the punishment of production. Besides being rugged, the systems must be simple and economical to run.

Air Gaging

Air gages sense by detecting changes in back pressure or flow. This type of noncontact sensor is becoming very popular again. When electronic sensors were introduced some years ago, the use of air gaging fell off. Early air gaging could not take readings and transmit them to another device for storage and later use. Now, air gaging is combined with electronics so data can be easily collected and analyzed in a computer.

An air-electronic transducer and analog-to-digital converter are used to collect and transmit data, increasing the stability and usefulness of the measuring process. The new air gaging measuring process is software driven, adding to its value and flexibility.

Air gaging regained popularity because it offers high resolution at low cost, can be adapted for high-volume operations easily, and is simple to maintain since it is self cleaning in nature.

Air gaging uses the restriction of airflow between a nozzle tip and the part being tested to learn part size. This type of gage needs a regulated air supply, a flow metering device, and one or more nozzles. A typical pneumatic circuit is shown in Figure 14-3.

In operation, the air from the regulated supply flows through the restriction and then through the nozzle. When the nozzle is free and open to the atmosphere, it will handle a maximum flow of air. In addition, there will be a minimum of pressure in the system downstream of the restriction. If a plate is moved in front of the nozzle and

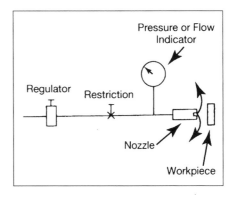

Figure 14-3. Basic components of an air gaging system (Edmunds Gages).

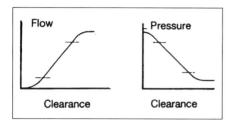

Figure 14-4. Relationship of airflow and clearance in an air gaging system (Edmunds Gages).

slowly brought toward it, the airflow will gradually be restricted until the nozzle is shut off. At this point, the airflow would be zero.

When the nozzle has been completely closed off, the pressure downstream of the restriction will build until it becomes the same as the regulated supply. Figure 14-4 shows a plot of both airflow and air pressure versus the distance between the plate and nozzle. Except for the extremes of flow and pressure, the curve is virtually a straight line. Because of this linearity, the distance of the plate from the nozzle can be determined.

Advantages. Using air gaging may have many advantages for a company. These are:

1. For measuring holes, air gaging is unsurpassed for speed and accuracy.
2. Small tolerances can be measured with air gaging since it has sufficient magnification and reliability.
3. An air gage can measure very small holes.
4. Concepts are simple. Operators do not require special training.
5. Air gaging is low in cost and can be readily modified for other applications.
6. Air gaging can be used for checking soft, highly polished, thin-walled, or otherwise delicate materials.
7. Air gaging cleans itself by the air escaping from the tooling. This reduces the possibility of false readings due to oil, dirt, or coolant on the part.

Back-pressure Gage System

Several types of back pressure gage systems have been developed (Figure 14-5). One type consists of air under constant pressure passing through an orifice of predetermined or adjustable size and into the gaging member. A pressure indicating device, or dimension indicator, is placed in the system between the orifice and the gaging member. This shows the change in pressure from the air escaping between the gaging member and the part.

A second type, called a water-column gage system, has constant pressure maintained by the height of the water column. The level of the liquid

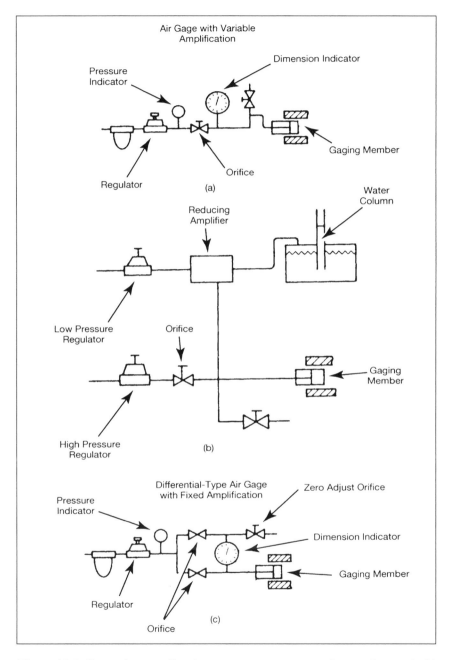

Figure 14-5. General types of back pressure gage systems that can be used with electronic data collection: (a) variable amplification, (b) water column, and (c) differential (Western Gage Corp.).

in the graduated tube shows the constriction facing the escaping air in the gaging member.

A third type uses a parallel circuit. Air at constant pressure enters two separate channels. The air in one channel is allowed to escape through the gaging member. In the other channel, air is allowed to escape through the zero adjust orifice. In operation, the air escaping between the gaging member and the part is compared with the amount escaping when a calibration master is in place. The pressure differential is registered by a dimension indicator and converted to an electrical signal using a pressure transducer.

Air Gage Tooling

Noncontact tooling, also called open-jet tooling, uses the direct flow of air from the air escapement orifice to contact the part. Rate of flow depends on the nozzle hole diameter and the clearance between the jet and part.

A noncontact gage head may have single jets, dual jets, or multiple jets. Because single-jet tooling is inherently sensitive to gage position, it is not used for inside and outside diameter applications. Dual-jet tooling is used in air probes for checking internal diameters. It also can be used in air snap and ring gages for checking outside diameters. This type of tooling eliminates the need for precise placement of the part. Figure 14-6 shows some dual-jet tooling applications. Out-of-round, bell mouth, hourglass, and barrel-shape conditions can be detected by rotating the gage. Air rings and snap gages with three jets are used for checking diameter, taper, and out-of-round conditions.

Laser Systems

In general, laser measurement systems can be viewed as projection systems. A laser beam is directed toward the part being checked so a spot from the laser is on the surface of the part. For scanning, the laser beam is repeatedly deflected so that a scan takes place with many spots rapidly occurring on the part surface. The part also may be moved in a set pattern so a sizable area of the surface is examined by the scan.

Light from each spot is reflected back to optical receptors on the laser system. Computer software analyzes the reflection and defines the shape and dimensions of what may be a very complex surface. Data can be fed into statistical process control packages to aid in controlling the process.

A laser scanning system generally requires a means of scanning, a high-spot density when scanning, and collectors that gather the light. The spot density defines the resolution of the measuring process.

Laser scanners are capable of much higher speed than vision measuring processes. This is because the velocity of the scan is normally very high

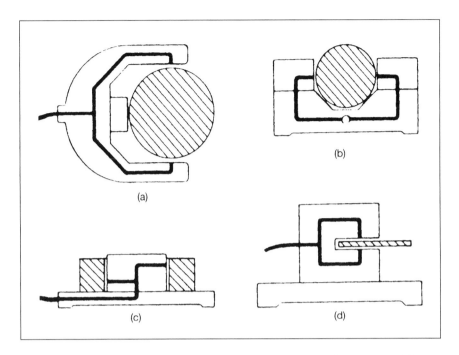

Figure 14-6. Examples of dual jet tooling used in an air gaging system: (a) snap gage, (b) V block gage, (c) perpendicular measurement, and (d) thickness measurement (Sheffield Measuring Div., Warner and Swasey).

and may likely produce hundreds of complete scans each second. Scanning is just one way a laser system can measure dimensions. Triangulation, often called a laser indicator, also may be used. This approach calls for the reflection angle of a beam to be bounced off the surface. The beam in a laser indicator is steady and continuous instead of scanning.

The word "laser" is an acronym that stands for "light amplification by stimulated emission of radiation." Laser light differs from ordinary light by being extremely intense, highly directional, strongly monochromatic, and coherent to a high degree. Laser beam intensity is the result of the light energy at a single wavelength in a particular direction.

Advantages

The advantages of using a laser system are:

1. No contact is required by the sensor and the part being measured.
2. The distance from the sensor to the part being measured can be large.
3. The response time is limited only by the speed of the photo detector or its electronics.

4. Light variations or interruptions are directly converted to electrical signals.

Scanning

High-speed measuring processes can use a scanning laser beam. The transmitter section of the gage emits a moving beam of light that scans at a regular speed. The object being measured interrupts the beam, and the detector determines the time that the beam took to traverse the part. The electronic controller converts the data into discrete dimensional readings for end use. This process is very stable.

Scanning Operation

A typical laser scanning instrument consists of a transmitter module, a receiver module, and processor electronics, as shown in Figure 14-7. The transmitter contains a low-powered gas laser, a power supply, a collimating lens, a multifaceted reflector prism, a synchronizing pulse photo detector, and a protective window.

In operation, the transmitter module produces a scanning laser beam, collimated and parallel, moving at a high steady linear speed. The scan-

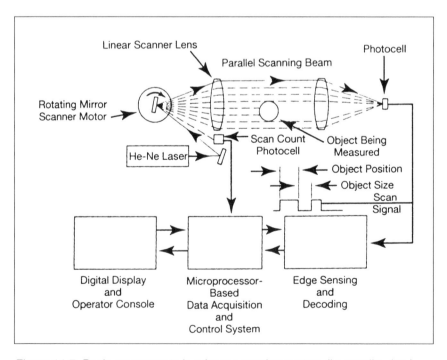

Figure 14-7. Basic components in a laser scanning system (Lasermike, Inc.).

ning beam appears as a line of red light and sweeps across its measurement field. A part placed in the field interrupts the beam. The receiver module collects and photoelectrically senses the laser light transmitted past the part being measured. The processor handles the signals from the receiver, electronically converting them to a usable form and then displays the dimension being measured.

Applications for Scanning

Laser scanning instruments can be used in a broad range of manufacturing operations and in a variety of industries. Some potential areas of application are wire manufacturing, centerless grinding, plastic extrusion, metal product fabrication, and nuclear reactor metrology. By modifying the techniques with which the detector output is digitized and interpreted by the processor unit, measurements can be made on translucent material such as fiberoptic cables or transparent material such as glass tubing. By examining the detector output in different ways, other measurements are possible, like product position and gap size.

More elaborate scanner geometries can be used to achieve dual-axis inspection. In these applications, the laser beam is alternately swept across the measurement field in two axes 90° apart. By stacking individual scanners back to back or detecting only the edge of a product and relating it to the position of a reference edge, parts much larger than the range of an individual scanner can be measured.

Extra-high-speed scanners, which measure at 4 to 6 times the normal rate, allow detection of smaller defects, such as lumps, in parts moving down a conveyor.

Triangulation

Several techniques have been used for range-type measurements. These include measuring the width of a spot with a conical or tapered beam, measurement of spacing between two projected spots with the input angles being selected for sensitivity, and single-spot laser triangulations.

Of the three techniques, laser triangulation is the preferred method for high-accuracy measurements. The simplicity of this method allows development of a highly rugged, reliable, and, for the most part, accurate device as part of the mechanical and optical hardware.

Triangulation Operation

A single-spot laser triangulation is shown in Figure 14-8. The sensors use a small, low-powered laser. A beam of laser light from the source is projected and focused by optics to the surface of the object. This creates a spot of light on the object surface similar to a flashlight beam on a wall. At some angle about the axis of the projected laser light, an image of the spot falls on the detector. As shown in Figure 14-8, the location of the centroid of the imaged spot is directly related to the standoff distance from the sensor to the object surface. A change in standoff distance results in a lateral shift of the spot centroid along the sensor array. The sensor processor calculates the measurement, which is frequently remotely mounted from the sensor head itself. The microprocessor-controlled sensor typically provides a readout of the surface standoff distance. A variety of single-spot laser triangulation sensors are available.

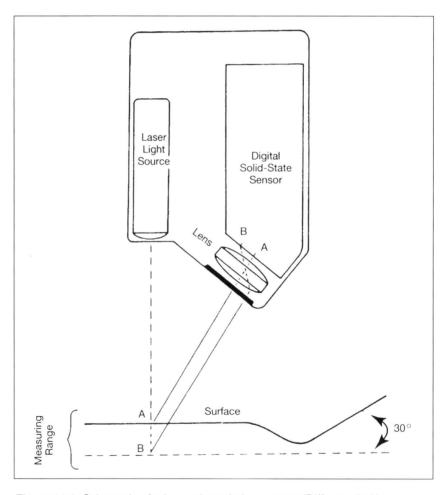

Figure 14-8. Schematic of a laser triangulation system (Diffracto, Ltd.).

Applications for Triangulation

Single-spot laser triangulation sensors have many applications in the industry. One such area is in dimensional gaging of components where the sensor provides a direct replacement for a contact (LVDT) or a noncontact probe. The advantage of such sensors over LVDTs includes remote measurement without contact and a high data rate. Concerning air and capacitance gaging techniques, laser triangulation permits longer standoff distances, higher response, and greater resolution.

Laser triangulation sensors are used as a replacement for the touch trigger probes typically used on coordinate measuring machines. In this application, the sensor determines surface features and locations using an edge-finding feature. However, the sensors cannot be used for probing down the depth of a bore to learn size or location. When used for surface measurement applications, the advantage of using a laser triangulation sensor on the coordinate measuring machine is a significant increase in measurement speed.

Machine Vision

With machine vision, parts are scanned and a video image is taken. Rather than measuring the part, this technology measures the image. Adjustments are made for the part orientation before an analysis is made.

Clear images of the part are assured by precision optics, proper illumination, and a high-resolution video camera. The video image is analyzed according to the instructions given by a computer software program.

A computer monitor can be used to view the part and its features. Also, the status of various part features (whether or not they meet the specification) is provided through computer software. Data can be fed into statistical process control packages to aid in controlling the process. Machine vision can readily magnify an image through a zoom option, making it the choice for measuring parts with small features that cannot be measured by contact sensors. In just five years, these systems have increased their speed more than three times.

To be classified as machine vision, the system must be capable of:
1. Forming the image, in which incoming light is received from a part and then converted into electrical signals.
2. Formatting the signals to be compatible with computer processing capabilities.
3. Analyzing and measuring various features or characteristics of these signals that represent the image.

4. Interpreting the data so that decisions can be made about the part being studied.

Advantages

The general advantages of vision systems are the same as previously mentioned for laser systems.

Operation of Machine Vision

The machine vision process consists of four basic steps:
1. An image of the part is formed.
2. The formed image is transformed into digital data that can be used by the computer.
3. The characteristics of the image are enhanced and analyzed.
4. The image is interpreted, conclusions are drawn, and action is taken.

Image Formation

Image formation frequently involves a combination of adjusting lenses, mirrors, and prisms that focuses the relevant portion of the part on the photo detector. The important parameters for image formation are lens focal length, aperture, depth of field, and magnification. Besides these parameters, the object must be properly illuminated to provide a quality image.

Illumination

Image formation begins by illuminating the part. The source of illumination must give the vision system the best possible images of the part. Images should have a good contrast between the features and their backgrounds to make an analysis possible. The lighting design should also ensure that extraneous lighting around the machine will not interfere with its operation. Light intensity, angle, wavelength, and structure are the major factors contributing to the success of an image processing installation.

Typical light sources used on machine vision systems include incandescent bulbs, fluorescent tubes, fiberoptics, arc lamps, and strobe lights. In addition, laser beams and x-ray tubes are used for special imaging applications. Occasionally, polarized or ultraviolet light also is used to reduce glare or increase contrast. Three basic lighting techniques are used for machine vision systems. They are front lighting, back lighting, and structured lighting, as shown in Figure 14-9.

When selecting a lighting technique for a particular application, the following factors should be considered:

1. What feature on the part is to be identified?
2. Is the part moving or stationary during inspection?
3. In what type of environment will the application be used?
4. What type of sensor will be used for the application?

Image Sensing

Sensing involves the formation of an optical image and the conversion of that image into electrical signals by some photosensitive target. When selecting an electronic imaging device for a particular application, several factors should be taken into consideration. These factors are field of view, resolution, contrast, signal-to-noise ratio, spectral response, dynamic range, geometric distortion, lag time and temperature stability, and cost.

Commercial machine vision systems use imaging devices or cameras that generate images as two-dimensional arrays or one-dimensional linear arrays. Two-dimensional arrays are similar to those images on conventional television cameras. One-dimensional linear arrays create the image by scanning the scene one line at a time. To capture a complete two-dimensional scene, the second dimension is obtained either by motion of the object past the scanner or by mirror or prism deflectors.

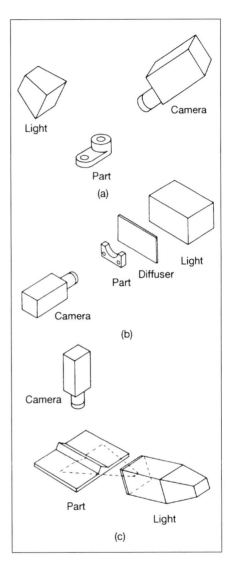

Figure 14-9. General types of lighting techniques used in machine vision: (a) front lighting, (b) back lighting, and (c) structured lighting (Penn Video).

The most commonly used camera in early vision systems was the Vidicon camera, which provided a great deal of information at high speed for relatively low cost. An image was formed by focusing the incoming light through a series of lenses onto the photo conductive faceplate of the Vidicon tube. Since then, much development of cameras has taken place. Now high resolution solid state color cameras appear to offer added advantages and should make vision systems more efficient. Research continues on high-definition video (HDTV) and other systems.

Image Analysis

The third general step in the vision sensing process is to analyze the digital image so that conclusions can be drawn and decisions made. This can be done in the system's computer. The image is analyzed by looking at the properties of the features in the image.

In general, machine vision systems begin the process of interpreting images by analyzing the simplest features. The analysis continues by adding more complicated features until the image is clearly identified. Many different techniques are being developed for vision systems. The general purpose of each approach is to analyze the image features describing the position and geometric construction of the part. Some simple geometric features may be area, perimeter, diameter, centroid, curvatures, or angles. More complex features may be edge gradients, spatial frequency content, projections, histograms, or convolutions.

Image Interpretation

When the system has completed the process of analyzing image features, some conclusions should be made about the findings. Some may be very elementary, such as whether the desired part is present and identified in the image.

Based on the conclusions, certain decisions can then be made about the part or the production process. These conclusions are formed by comparing the results of the analysis with the expected results established by a preconceived computer model of the image. Alternately, a model could be built by sampling previously examined items.

In cases where several image features must be measured to interpret an image, a simple factor weighting method may be used to consider the relative contribution of each feature to the analysis. For example, to identify a valve stem from among a group of stems of several sizes, the image area may not be sufficient by itself to ensure a positive identification. The measurement of height may add some additional information. Features known to be the most likely indicators of a match would be weighted more than others. A weighted total goodness-of-fit score is found. The score suggests the likelihood that the object has been successfully identified.

Template Matching

In template matching, a mask is electronically generated. The template must match the image of an object. When the system checks objects to recognize them, it matches "the image of" with the available templates. With a perfect match, all pixels would align perfectly. If the objects are not precisely the same, some pixels will fall outside the standard image. The percentage of pixels in the two images that match is a measure of the goodness of fit. The confidence that a correct interpretation has been made is calculated.

Performance Characteristics

Besides understanding how machine vision works, it is also useful to understand the basic criteria by which machine vision systems can be evaluated and compared. These represent the basic considerations that potential users should review when deciding about vision systems. In general, the ideal system is one that allows fast, accurate interpretation of many complex images with a minimum of jigs and fixturing.

Resolution. The ability of a vision system to create a recognizable image of an object is determined by the number of pixels in the image. Much progress has been made in this area. Currently, available systems have a million pixels in an image.

Processing Speed. A machine vision system could form an image, analyze, and interpret it at a speed consistent with the speed at which parts are being presented. Two categories of speed can be considered.

First, image processing speed measures the number of bits of information that can be processed by the image processor. The speed at which individual items can be examined by the system may be more important. This is a difficult number to find because processing time is affected by many factors, including complexity of the image, type of illumination, and accuracy required in interpreting the image.

Discrimination. The ability of a vision system to discriminate between variations in light intensity over an image is determined by the number of intensity thresholds present in the system. Early installations used a binary system, which allowed two levels of intensity. This provided the least ability to discriminate. Much development has taken place since then. Now a high-resolution solid state color system can perceive more subtle intensity variations.

Accuracy. A tradeoff can be made between processing speeds and the ability to interpret images correctly. A higher probability of correct interpretation can be achieved by processing more image features that, of course, increase the processing time.

Accuracy, also known as consistency, can be defined as the percentage of correct decisions made about a group of objects being examined by a vision system. This is a difficult number to estimate because it depends on such things as illumination and how much programming was done, as well as many other factors. An acceptable accuracy rate depends on the precision required by the application.

Applications of Machine Vision

The following are inspection applications for machine vision:
1. Gaging—checking to make sure that dimensions are within acceptable tolerances.
2. Verification—checking to make sure that a product is present, complete, or the right one in the proper orientation.
3. Flaw detection—checking for unwanted features anywhere on the observed portion of the product.

COORDINATE MEASURING MACHINES

Coordinate measuring machines (CMMs) that provide three-dimensional measurement, are owned by many companies. The vast majority of these companies consider the machines critical to continuous process improvement. These companies need measurement data quickly to avoid making substandard parts. A testimonial to their importance to manufacturing is that more than one half of the CMMs produced so far are used in a production setting.

Critics feel that CMMs have lagged behind machine tool development because CMMs were isolated in quality departments in the past. Lending support to this viewpoint is that many improvements have taken place since manufacturing has assumed much of the quality responsibility.

For example, CMMs are quickly catching up with the long available technology of CNC machines. With the help of interfacing schemes, a CAD model can now be electronically transmitted to the CMM. Using this interface, probe paths can be more easily defined to tell the CMM how to measure the features on a part.

This progress is continuing with the adoption of the Dimensional Measuring Interface Standard (DMIS) as a national standard. Now, CMMs rely largely on contact probes or sensors. Many feel that noncontact sensors need to be used even more for CMMs.

Laser-based sensors on CMMs have been successful in measuring vehicle dimensions on production lines and have played a big role in helping to get better quality. More use in applications like this will translate into higher operating speeds, as well as greater accuracy and precision.

Further improvements for CMMs are felt by some professionals to be the biggest challenge facing manufacturing in the near term. Additional progress hinges on full integration of CMMs into the manufacturing setting and, in particular, the business and manufacturing processes.

Since their development, CMMs have been increasingly used throughout industry. However, they are most predominant in the automotive and aerospace industries. Although they were once considered an exotic tool for ensuring quality control, CMMs are now becoming a mandatory piece of equipment for both the large manufacturing plant and the small job shop. This is due primarily to the need for an accurate measuring instrument and the need for detailed documentation of the components being produced.

CMM Placement

Currently, CMMs are used in one of three ways. The simplest approach is to place the CMM at the end of the production line or in an inspection area. Here, the CMM inspects the first part of a production run to verify the machine setup and then proceeds to measure parts on a random basis.

Another approach is to use CMMs between two work centers. Parts produced at the first center are measured before any secondary operations are performed. This approach is possible because CMMs can make many different measurements within a short period. With this method, the CMM indirectly controls the production process.

A third approach places the CMM in the production line, permitting the machine to directly control the production process. In operation, a CMM first measures the part, then a software program compares the measurements with required dimensions. If necessary, process control devices adjust the machine so that the part is manufactured within the required specifications.

CMM Elements

A CMM consists of four elements. The mechanical structure of the CMM is basically a positioning device using X-Y-Z axes. The probing system collects raw data on the part and provides input to the control system. Then there is the machine control and computer hardware. And finally, there is the software for three-dimensional geometry analysis.

The measuring envelope is defined by the X, Y, and Z axis travel. Although a variety of machine designs and configurations exist, all designs use three coordinate axes. Each axis is perpendicular to the plane formed by the other two axes, and fitted with a linear measurement transducer for positional feedback.

The most commonly used scales in CMMs are stainless steel and glass. Both rely on noncontact electro-optical reader heads for determining the exact position of the CMM probe. Stainless steel scales are widely used in shop environments. This is because most parts are made from steel and measurements will not be off by much because the coefficient of thermal expansion of the scale and a steel part are very close.

Glass scales are generally used in controlled environments. There is a large difference in the coefficient of thermal expansion between glass and parts made of most other materials. The use of temperature compensation in some CMMs reduces differences in measurements due to temperature.

While scales used in CMMs are either glass or steel, parts checked could be of many different materials, and a correction must be made. Complicating this problem is that the published coefficients for materials have only a 10% accuracy.

The work table of a CMM usually contains tapped holes to ease the clamping and locating of parts. It may be made from granite because of its stability in various environments. Electronic or solid probes are inserted into the probe arm. Probe arm movement is guided by means of frictionless air or mechanical bearings.

CMM Measurement Process

Coordinate measurement is a two- or three-dimensional process that determines the position of holes, surfaces, center lines, and slopes. All six sides of a cube-shaped part may be inspected without repositioning.

In a typical operation, the part is placed on the table of the CMM at a random location. Generally, the part is centered so that all part surfaces can be inspected with the probe. Depending on the size of the part and the type of probes, the part may need to be clamped to the machine table. If several checks of similar parts are required, a simple fixture may be used. The probe is then moved, manually or by machine control, until contact is made with desired part features. Reader heads, on each axis, transfer the position to the computer interface. The dimensional and geometric elements may then be calculated, compared, and evaluated, stored, or printed out as required.

Advantages

Why should an organization invest in one or more CMMs? The technology offers a number of advantages.

Flexibility

CMMs are essentially universal measuring machines and do not need to be dedicated to any single or particular measuring task. They can measure practically any dimension on virtually any part.

Reduced Setup Time

Procedures for establishing part alignment and appropriate reference points are greatly simplified on CMMs. In contrast, they are often quite complex and time consuming for conventional surface plate measurements. Software allows the operator to define the part's orientation on the CMM. All coordinate data are corrected for any misalignment between the part and the machine. A CMM can inspect complete parts in a single setup without the need to reorient the part.

Improved Accuracy

All measurements on a CMM are taken with a common fixed measuring system. This reduces errors that can result with layout measuring methods. Moreover, measuring a part completely in one setup prevents the introduction of errors when changing the setup.

Reduced Operator Influence

The use of digital readouts and computer printouts eliminates subjective interpretation of readings common with dial or vernier-type measuring devices. Operator "feel" is virtually eliminated with modern electronic probe systems. In computer controlled CMMs, machines may not need the undivided attention of operators. Also, less skilled operators can be readily instructed to operate the CMM effectively.

Improved Productivity

All factors mentioned help make CMMs more productive than conventional inspection techniques. Further dramatic productivity improvements are realized through the computational and recording ability of the software.

Types of CMMs

A variety of machine configurations are available. Each configuration has advantages that make it more suitable for a particular application.

Cantilever

Cantilever CMMs have three movable components moving along mutually perpendicular guideways. A typical machine of this configuration is

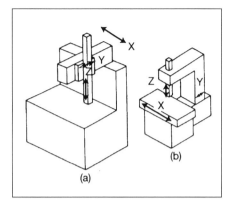

Figure 14-10. Examples of cantilever CMMs: (a) fixed table and (b) moving table (American Society of Mechanical Engineers).

shown in Figure 14-10. A modification of the fixed table cantilever configuration is the moving-table cantilever CMM, also shown in Figure 14-10. Cantilever-type CMMs are usually the smallest and lowest in cost and occupy a minimum of floor space. This configuration permits a completely unobstructed work area, allowing full accessibility to load, inspect, and unload parts that may be larger than the table.

Bridge

Bridge CMMs also have three components moving along mutually perpendicular guideways. A typical machine of this configuration is shown in Figure 14-11. This type of CMM is often called a moving bridge or a traveling bridge CMM.

A modification of the moving-bridge configuration has each end of the bridge structure fixed to the machine base, as shown in Figure 14-11. The part is mounted on a separate table that moves horizontally. This configuration is called a fixed-bridge CMM.

Another type of bridge configuration shown in Figure 14-11 has two bridge-shaped components. One of these bridges is fixed at each end to the machine base. The other bridge, which is an inverted L-shape, moves horizontally on guideways in the fixed bridge and machine base.

A third type of moving bridge configuration is the central bridge, also shown in Figure 14-11. The drive forces are applied to the center of the bridge. This reduces the bridge pitching and yawing when moving, allowing higher acceleration and deceleration rates.

The bridge CMM is the most popular configuration. Its double-sided design provides more support, which is needed in large- and medium-sized machines. The bridge can slide back on the base to give complete accessibility to the working area for easy loading and unloading of parts.

Column

Column CMMs are similar in construction to jig boring machines, as shown in Figure 14-12. The column moves in a vertical direction and a two-axis saddle permits movement in the horizontal direction. Column

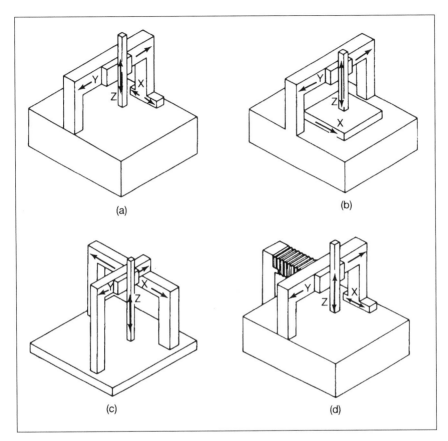

Figure 14-11. Examples of bridge CMMs: (a) moving bridge, (b) fixed bridge, (c) L-shaped bridge, and (d) central drive bridge (American Society of Mechanical Engineers).

CMMs are often called universal measuring machines rather than CMMs by manufacturers. They are considered gage-room instruments rather than production floor machines.

Gantry

Gantry CMMs have three components moving along mutually perpendicular guideways, as shown in Figure 14-13. The probe moves vertically about a cross beam, which moves along two elevated rails supported by columns attached to the floor. The gantry-type configuration was made initially to inspect large parts. These include such things as airplane fuselages, automobile bodies, ship propellers, and diesel engine blocks. The

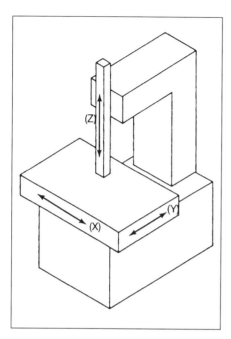

Figure 14-12. Example of a column CMM (American Society of Mechanical Engineers).

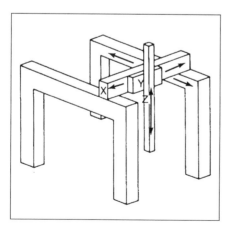

Figure 14-13. Example of a gantry CMM (American Society of Mechanical Engineers).

open design permits the operator to remain close to the part being inspected.

Horizontal Arm

Several different horizontal arm CMMs are available. Like all CMMs, the horizontal arm configuration has three movable components moving along guideways. In the moving ram CMM, the probe is attached to the horizontal arm. The ram is encased in a carriage that moves in a vertical direction and is supported on a column that moves horizontally. In the moving table CMM, the probe is also attached to the horizontal arm, which is permanently attached at one end to a carriage that moves in a vertical direction on the column. The arm support and table move horizontally about the machine base. For the fixed table CMM, the probe also is attached to the horizontal arm. A cantilever design is used to support the probe as it moves in a vertical direction. The arm support moves horizontally about the machine base. Several types of horizontal arm CMMs are illustrated in Figure 14-14.

Horizontal arm CMMs are used to inspect the dimensions of a broad spectrum of machined or fabricated workpieces. Parts to be inspected are mounted on the machine base. Using an electronic probe, these machines check parts in a mode similar to the way they

are machined on horizontal machine tools. They are especially
suited for measuring large gear
cases and engine blocks. By using
a rotary table, four-axis capability
can be gained.

Probes

The usefulness of a CMM depends largely on the nature of the
probing device. Three types of
probes are commonly used:

- hard contact;
- touch; and
- noncontact.

A probe is usually chosen based
on the dimensional and geometri-
cal requirements of the measuring
process.

Hard Contact Probes

Hard contact probes have been
used on CMMs since the mid-70s.

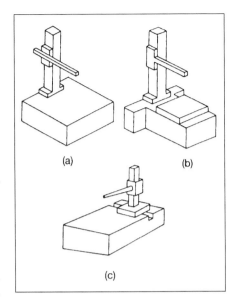

Figure 14-14. Examples of horizontal arm
CMMs: (a) moving ram, (b) moving table,
and (c) fixed table (American Society of
Mechanical Engineers).

They are still used on those machines that are not computer controlled. An
operator must manually position the probe on the part before the measuring coordinates of the probe can be taken.

The hard contact probe consists generally of a probe tip attached to a
probe arm. A variety of probe tip shapes and sizes are available. The shape
of the tip determines its application: conical tips are used for locating holes;
ball probe tips are for establishing surface locations; cylindrical probe tips
are for checking slots and holes in sheet metal parts; and edge finder probe
tips are used for part alignment and measurement of flat surfaces or edges
of parts.

Hard contact probes should be used only in small, manually operated
CMMs for inspecting simple parts. Being phased out, they still are widely
used on manual CMMs. Unfortunately, this type of probe allows much
operator error to enter the measurement. Each operator has his or her own
"feel" when contact is made by the probe on the part. Like micrometers,
there could be much variation in these measurements due to the subjective
judgment of the operator.

Inspection and Measurement

Touch Probes

Development of better touch probes continues. They are absolutely necessary in computer operated CMMs. Requirements for more intensive checking may result in more than five million measurements taken on one CMM annually. Without programmed checking paths and probes to support it, this would not be possible. Currently, most touch probes are based on a kinematic/resistive technology. The design of these probes uses a three-point kinematic reseating mechanism that is highly repeatable. This mechanism restores the stylus ball of the probe to its original position very quickly after measuring.

When the part is contacted by the probe, the kinematic mechanism is unseated and the stylus ball is depressed slightly. The resistance changes in the probe and a signal triggers the CMM computer to start recording the probe's position. At this point, the measurement is completed and the probe is withdrawn and sent to the next point on its programmed path.

Unfortunately, with this type probe there is some variation in how much the stylus ball travels before sending the trigger signal. Much of this variation is due to variation in the stylus bending at different approach angles. This situation, called lobing, is more pronounced with a longer stylus and when high trigger forces are used. Some probe designs can eliminate lobing. One such design uses the same kinematic mechanism for returning the stylus ball to its original position. But, the new design isolates the measuring sensor from the kinematic mechanism, eliminating lobing. Stylus bending is almost constant, despite what direction the probe approaches.

Noncontact Probes

Development of noncontact probes also continues and improvements are being seen. Noncontact probes are used when fast, accurate measurements are required with no physical contact with the part. Several different types of noncontact probes may be used, including those previously discussed in this chapter.

Optical Probes

Optical probes are useful for checking small, fragile workpieces. The two types of optical probes still used on manual CMMs are the projection microscope and centering microscope.

With the projection microscope, the image under inspection is displayed on the screen. Part feature locations are obtained by moving the CMM to align the screen reticle to the feature. With the centering microscope, part feature locations are obtained in the same way as when the user looks through the eyepiece. A third type of noncontact probe contains a laser light source that projects a small diameter spot on the part surface. A digital solid-state sensor detects the position of this spot and computes part surface location by optical triangulation.

Other types of noncontact sensors are also available for all CMMs. Development of new applications for other types of noncontact probes continues for computer controlled CMMs. At this time, touch probes are extensively used for measuring with computer-controlled CMMs.

Environment

Environment is critical to the performance of any CMM. The effect of temperature was discussed earlier in this chapter. Vibrations also can affect CMM accuracy, including both continuous vibration and random shocks. Often, a CMM is installed with vibration isolators in place for control of vibration.

Airborne particles also can cause deterioration. Dirt that is wet in nature will tend to stick and clog the machine, causing its performance to slip. Components should be shielded from dirt, grease, and other contamination.

A CMM installed in a hostile environment must be rigged to withstand the rigors of the environment. A CMM designed for the laboratory cannot be expected to last long in a production setting.

A well-designed preventive maintenance program will help a company protect its investment and assure that a CMM performs well when needed.

Bibliography

Alderman, B. Wendall, "The Evolution of Column Gages." *Quality Progress*, November 1992, p. 69.

Aronson, Robert B., "Inspection and Quality Assurance." *Manufacturing Engineering*, August 1994, p. 184.

Aronson, Robert B., "The Growing Importance of Gaging." *Manufacturing Engineering*, April 1995, p. 45.

"Automatic Measurement on an Optical Comparator." *Quality*, August 1993, p. 63.

Bosch, John A., "The Future for CMMs." *Quality*, August 1993, p. 57.

Busch, Ted, *Fundamentals of Dimensional Measurement*. Albany, NY: Delmar, 1989.

Ceglarek, Dariusz and Shi, Jianjun, "Dimensional Variation Reduction for Automotive/Body Assembly." *Manufacturing Review,* V8 N2, June 1995, p. 139.

Everhart, John, "Use of CMMs as Flexible Gages on the Production Floor." Dearborn, MI: SME *GAGETECH*, 1994.

Farago, Francis T. and Curtis, Mark A., *Handbook of Dimensional Measurement, 3rd ed.* New York: Industrial Press, 1994.

Freyberg, Dale and Traynor, Dennis, "How Do Optical Gages Work?" *Quality,* August 1991, p.11-Q.

Fundamentals of Tool Design, 3rd ed. Dearborn, MI: Society of Manufacturing Engineers, 1991.

Gazdag, William S., "High Accuracy CMMs." *Quality*, December 1992, p. 20.

Gazdag, William, "Micrometers Keep Getting Better with Time." *Quality Progress*, July 1990, p. 64.

Gazdag, William S., "The Use, Mis-use, and Non-use of CMMs." *Quality*, August 1993, p. 24.

Gazdag, William S. and Placek, Chester, "The Brave New World of Coordinate Metrology." *Quality*, August 1993, p. 21.

Genest, David H., "Coordinate Measuring Machines," reprinted from ASM Handbook Vol. 17: *Nondestructive Evaluation and Quality Control.*

Gilman, William R., "Measuring with Light." *Quality,* August 1993, p. 47.

Grove, John W., editor, *Handbook of Industrial Metrology.* Englewood Cliffs, NJ: Prentice Hall, 1967.

Harding, Kevin G., "Sensors for the '90s." *Manufacturing Engineering*, April 1991, p. 57.

Koelsch, James R., "Bench Top CMMs the PCs of Metrology." *Manufacturing Engineering*, April 1992, p. 49.

Krejci, James V., "Capable CMMs: A Buyer's Guide." *Manufacturing Engineering,* April 1993, p. 77.

McCarthy, William F., "The State of Noncontact Measurement Technology." *Quality*, August 1991, p. 5-Q.

McMurtry, David R., "The Future of CMM Probing Systems." *Quality,* October 1991, p. 21.

"Measuring and Gaging in the Machine Shop." Ft. Washington, PA: National Tooling and Machining Association, 1981.

Morris, Alan S., *Measurement and Calibration for Quality Assurance.* Hertfordshire, England: Prentice Hall, 1991.

Moyer, Mike, "Measuring Angular Errors." *Quality*, August 1991, p. 15-Q.

Nashman, M., "Vision and Touch Sensors for Dimensional Inspection." *Manufacturing Review*, V6 N2, June 1993, p. 155.

Noaker, Paula M., "Scrutinizing Surface Measurement." *Manufacturing Engineering,* April 1991, p. 47.

Owen, Jean V., "Inspection/Quality Assurance." *Manufacturing Engineering*, August 1990, p. 162.

Owen, Jean V., "Seeing the Unseen." *Manufacturing Engineering*, April 1993, p. 39.

Parlee, Kenneth, "Optical Comparators." *Quality,* August 1992, p. 23.

Parlee, Kenneth, "Role for Noncontact Sensors Growing." *Quality*, August 1993, p. 41.

Placek, Chester, "CNC/DCC Models Boost Benchtop CMM Sales." *Quality,* June 1992, p. 47.

Polidor, Edward T., "Noncontact: Faster Measurement Data." *Quality*, August 1991, p. 3-Q.

Rzeznik, Tom, "In-line Laser Measurement in the Assembly Process." Dearborn, MI: SME *GAGETECH*, 1994.

Sagar, Paul, "Automatic Temperature Compensating Gages." *Quality*, February 1992, p. 27.

Shelton, Russ, "Future Facts in Coordinate Metrology." *Quality*, August 1993, p. 53.

South, Veeder III, "A Case for Semiautomatic Optical Measurement." *Quality*, August 1992, p. 29.

Suntag, Charles, *Inspection and Inspection Management*. Milwaukee, WI: *ASQC Quality Press*, 1993.

Traylor, Alan, "Probes Keep Pace with DCC CMMs." *Quality*, August 1993, p. 36.

Valliant, J. Glenn, "A New Quantitative Tool for Quantitative Surface Analysis." *Quality,* October 1992, p. 35.

Wahi, Anita K. and Howland, Rebecca S., "Quantitative Surface Roughness Measurements." *Quality*, October 1992, p. 47.

Watts, William A. and Prout, L. Stephen, "The Need for Interim CMM Checks." *Quality,* October 1991, p. 25.

Wenzler, Christopher M., "Gages Help Many through Thick and Thin." *Quality Progress,* September 1990, p. 83.

Wick, Charles, editor-in-chief and Veilleux, Raymond F., staff editor, *Tool and Manufacturing Engineers Handbook, Volume 4, 4th ed.* Dearborn, MI: Society of Manufacturing Engineers, 1987.

Wolak, Jerry, "Gaging or Not." *Quality*, February 1992, p. 27.

Summary

Both inspection and measurement are alive and well. They are in safe hands—those of manufacturing professionals. It has only been a relatively short time since these functions were transferred from the quality department to manufacturing. Yet steady improvements are being seen.

Largely due to up-front planning by product development teams, inspection and measurement functions are being integrated into production processes. No longer standalone, they add value to manufacturing processes and contribute to a company's continuous improvement. Inspection, besides deciding whether a product meets needs, drives process control effort to make products better. Measurement instruments, particularly CMMs, are improving. Ongoing effort continues to bring measuring processes more in line with GD&T principles.

Inspection and measurement are needed because of variability. Any process produces a product for which the features and dimensions inherently vary. Each time the same product is inspected or measured, different results are probable.

Causes of this variability may be lack of:
- proper control of the process;
- uniformity of material or parts entering the process;
- control of the environment such as temperature, humidity, or vibrations; or
- control of human errors.

Inspection will continue to be key for processes in the future. But, it will be different from today for most companies. There will be less dependence on humans, and more on automated inspection. Sorting good products from those that are unacceptable will still be necessary to satisfy customers and internal needs. This may be critical on those features having severe safety or customer satisfaction implications. Inspection will be largely "in-line," "upstream," and integrated in the process to provide information to drive process controls. Variation in products will be reduced and may not be a major issue with improved process controls. Planning for inspection will start early in the new product development cycle. It also will be done concurrently with planning the process.

Planning for quality activities is largely based on the experience and judgment of those involved. To organize this effort, a control plan is prepared early in the product development cycle by a multifunctional team. The objective of a control plan is to help assure that all outputs of the manufacturing process are in control. It contains the actions required at the various stages of the process. During production, the control plan specifies the process monitoring and control schemes that will reduce variation of both process and product characteristics.

The desired practices for inspection and measurement activities are part of generic quality standards. The first widespread use of voluntary generic quality system standards was ISO 9000. The international standards were first issued in 1987 and the "teeth" that forced compliance was registration, done by third parties called registrars. Companies accept the results of these registrars usually without conducting audits themselves. Registration assures companies that a supplier has an adequate quality system that is periodically monitored.

There are guidelines for inspection and measurement that are very helpful to those planning quality activities. Requirements for the Malcolm Baldrige National Quality Award, as well as industry specific documents like QS 9000, describe activities needed in a company to go beyond an ISO 9000 quality system. For some companies, this may be helpful in getting a vision of the future. For others, it may be helpful for fine tuning their quality improvement efforts in the short term.

For most companies, doing tasks associated with inspection and measurement in a manual fashion is no longer feasible. Computers and software that meet the needs of those using the system are absolutely necessary. Computers make it less likely that something important will slip through the cracks.

Measurement is a process similar to any other process. It can be viewed as a set of interrelated resources, activities, and influences that produce a measurement. There are influences such as the environment that may add to the variability of the process.

The measurement process has inputs, outputs, and steps. In simplistic terms, inputs could be viewed as the article to be measured, the operator, procedures, and an instrument to do the measurement. One also could include as inputs reference standards for instrument calibration and a proper environment for doing the measurement. The output is, of course, the measurement.

Equipment for normal measurement consists of instruments, auxiliary apparatus, standards for comparison, reference materials, and the procedure or work instructions to do the measurement. Humans involved with making the equipment work also may be included as part of this structure.

To improve planning for measurement during product development, some companies in the automotive industry adopted a process that could be called dimensional assurance. It does not substitute for quality planning during product development, but dimensional assurance is superimposed as part of it. Dimensional assurance increases the likelihood that a good measurement system will be developed. Its principles are general enough to be adopted by any company as a separately named effort or incorporated into the quality planning activities of the product development cycle.

The broad acceptance by the professional community that measurement is a process recently led to a new approach for monitoring measurement systems. Recognizing it to be a process allows measurement to be monitored and controlled like other processes. Control charts, long used in manufacturing, can be used to monitor measurements on a real time basis. Sudden or gradual deterioration of the quality in measurements can be detected. The deterioration could be either from an increase in the random error or the systematic error, or both in combination.

A formal program is essential to assure the continued accuracy of instruments and measuring processes throughout the production life cycle. Often called a confirmation system, it is aimed at preventing inordinate measuring errors. A confirmation system does this by detecting errors before they are excessive and taking prompt action to correct them. In other words, the purpose of a confirmation system is to assure that the errors in measurement stay within acceptable limits. To accomplish this, each instrument or measuring process is checked periodically according to a predetermined schedule.

Certain measuring tools and equipment appear to dominate plans to achieve better measurement strategies for manufacturing. These include noncontact sensors and CMMs. This does not imply that the devices left out are unimportant to manufacturing. In reality, the opposite is true. Simple instruments, like micrometers and calipers, are basic and will continue to be important to manufacturing in the future. Likewise, contact sensors are widely used by most companies. These devices continue to make a valuable contribution to good manufacturing practices.

Inspection and Measurement

Many consider temperature to be the greatest cause of error when checking dimensions in a manufacturing setting. Sometimes just the heat of the operator's hand can introduce error. From a practical viewpoint, the integrity of measuring hinges on controlling temperature variations or, alternately, being able to make valid corrections for the differences.

The role in measurement for noncontact sensors is growing. Reduced tolerances and the continued pressure for lower costs drive the search for more efficient measurement strategies. This translates into those measuring instruments and processes that can provide both greater speed and accuracy.

Today, many companies have CMMs, with most companies considering the machines critical to the continuous improvement of their processes. These companies need measurement data quickly to avoid making substandard parts. A testimonial to their importance in manufacturing is that more than one half of the CMMs produced so far are being used in a production setting.

The future looks bright for inspection and measurement. There are many challenges that can be met by those in manufacturing, placing more and more emphasis on inspection and measurement.

Index